Gesellschaft für
Reaktorsicherheit (GRS) mbH

Safety Assessment of Unit 5 (WWER-440/W-213) of the Greifswald Nuclear Power Station

Common German-Soviet Report

Gesellschaft für Reaktorsicherheit mbH
together with
Kurchatov-Institute for Nuclear Energy
OKB Gidropress and Atomenergoprojekt

GRS-92 (February 1992)
ISBN 3 - 923875 - 42 - 8

Annotation

This report represents the translation of the documentation "Sicherheitsbeurteilung des Kernkraftwerks Greifswald, Block 5 (WWER-440/W-213), Gemeinsamer deutsch-sowjetischer Bericht", GRS report GRS-88.

In cases of doubt report no. GRS-88 is the factually correct version.

Keywords

PWR, operational experience, cooling system, safety device, reactor safety, material problems, confinement system, instrumentation and control, spreading impacts, radiation protection, external impacts, fuel rod behavior

Preface

The German Federal Ministry for the Environment, Nature Conservation, and Reactor Safety (BMU) has commissioned the Gesellschaft für Reaktorsicherheit (GRS) mbH to assess the safety of Unit 5 of the Greifswald Nuclear Power Station. Within the program of German-Soviet cooperation in reactor safety and radiation protection (Item 2, WWER Safety Analyses), this safety assessment was conducted jointly with Soviet experts. On the Soviet side, the following institutions participated in the safety assessment: Minatomenergoprom of the USSR (Ministry for Atomic Energy Industry), Gospromatomenergonadsor of the USSR (State Supervisory Committee), the Kurchatov Institute for Atomic Energy, Atomenergoprojekt, OKB Gidropress (chief designer), and the All-Union Institute for Nuclear Power Plants.

In addition, the investigations were conducted in close cooperation with the French Institut de Protection et de Sûreté Nucléaire (IPSN).

The opinions expressed by the experts from the USSR and the Federal Republic of Germany were discussed and minuted at joint meetings attended also by representatives of IPSN.

The basic approach proposed by the experts to upgrade Unit 5 of the Greifswald Nuclear Power Station, and the choice of the most important compensatory backfitting measures, are in agreement. A number of technical questions contained in the report still require additional investigations and analyses to be conducted. To clarify these issues, further joint working sessions have been scheduled by the German, Soviet, and French partners.

Provided the upgrading measures proposed in the report are implemented, no decisive conceptual defects are apparent at the present status of investigations which would, from the technical point of view, fundamentally jeopardize the plant commissioning and power operation.

CONTENTS

1	Introduction	1
2	Technical Plant and Systems Features of type WWER-440/W-213 Nuclear Power Plants	5
2.1	Primary System	6
2.2	Secondary System	7
2.3	Cooling Water Systems	10
2.4	Engineered Safeguards Design	10
2.4.1	Emergency Core Cooling and Residual Heat Removal System	11
2.4.2	Emergency Feedwater System	12
2.4.3	Service Water System and Component Cooling System	13
2.4.4	Confinement System	14
2.4.5	Ventilation Systems	15
2.4.6	Electric Power Supply	16
2.4.7	Instrumentation and Control	17
2.5	Figures, Section 2	20
	Table 2-1	35
3	Basic Legal Licensing Principles	37
3.1	Licensing Situation, Unit 5	37
3.2	Valid Legal Principles of Licensing	38
	References, Sec. 3	42
4	Reactor Core and Pressurized Components	43
4.1	Core Design	43
4.1.1	Neutron Physics Core Design	43
4.1.2	Thermohydraulic Core Design	45
4.1.3	Mechanical Design of Reactor Pressure Vessel Internals and of the Core	46
4.1.4	Fuel Assembly Damage during Handling	49
4.2	Pressurized Components in the Primary and Secondary Systems	50
4.2.1	Purposes	50

4.2.2	Safety-related Assessment and Measures Required	51
5	**Loads Resulting from Accidents**	**59**
5.1	Analyses of Loss-of-Coolant Accidents and Transients	59
5.1.1	Loss-of-Coolant Accidents	60
5.1.2	Transients	62
5.1.3	Cold Water Strands	64
5.2	Pressure-resistant Compartment System with Pool-Type Pressure Suppression System, so-called Confinement System	65
5.2.1	Basic Project Design Principles	65
5.2.2	Analysis of the Design Parameters of the Confinement System	66
5.2.3	Differential Pressure Loads	67
5.2.4	Dynamic Loads Acting on the Pool-Type Pressure Suppression System during Accidents	67
5.2.5	Jet Forces and Reaction Forces	68
5.2.6	Leaktightness and Confinement Isolation	68
5.2.7	Summary Assessment and Measures Required	69
5.3	Radiological Impacts	71
5.3.1	Loss-of-Coolant Accidents	72
5.3.2	Fuel Assembly Damage during Handling	73
5.3.3	Break of the Steam Generator Collector	74
	References, Section 5	74
6	**Systems Engineering**	**76**
6.1	Systems Analysis of Process Engineering Aspects	76
6.1.1	Initiating Events	76
6.1.2	Event Sequences in Loss-of-Coolant Accidents	78
6.1.2.1	Large Leak (NB 200 to NB 500)	78
6.1.2.2	Medium-Sized Leak (NB 25 to NB 200)	80
6.1.2.3	Small Leak (<NB 25)	81
6.1.2.4	Pressurizer Leak	82
6.1.2.5	Leakage of a Steam Generator Tube	83
6.1.2.6	Leakage of Several Steam Generator Tubes or Leakage of the Steam Generator Collector	84

6.1.2.7	Leakage in a Pipe Connecting to the Primary System outside the Confinement System	85
6.1.3	Sequences of Transient Events	85
6.1.3.1	Loss of Main Heat Sink	85
6.1.3.2	Loss of the Main Feedwater	86
6.1.3.3	Failure of Turbo Generators	87
6.1.3.4	Leak of a Main Steam Line	87
6.1.3.5	Leak of the Main Steam Collector	88
6.1.3.6	Leak of a Feedwater Pipe	88
6.1.3.7	Leak of a Feedwater Collector	89
6.1.3.8	Failure of Main Cooling Water System and Service Water Systems	90
6.1.3.9	Startup and Shutdown Events	91
6.1.3.10	ATWS Accidents	91
6.1.4	Summary	92
6.2	Electric Power Supply	93
6.3	Instrumentation and Control	95
6.4	Ergonomics	97
	References, Section 6	99
7	**Spreading Impacts, Civil Engineering Aspects, Radiation Protection**	**100**
7.1	Spreading Impacts	100
7.1.1	Yardsticks Used for Reference	100
7.1.2	Spreading In-Plant Events	101
7.1.2.1	Fire	101
7.1.2.2	Flooding	105
7.1.2.3	Other In-Plant Spreading Impacts	108
7.1.3	External Impacts	108
7.2	Civil Engineering Aspects	109
7.3	Radiation Protection	112
7.3.1	External Impacts of Normal Plant Operation	112
7.3.2	Radiological Protection of Workers	113
	References, Section 7	115
8	**Evaluation of Operating Experience**	**116**
8.1	Work Performed	116

8.2	Upgrading Measures Required	118
8.2.1	Mechanical Systems	118
8.2.2	Instrumentation and Control	119
8.2.3	Station Service Power Supply	121
8.2.4	Building Structures	121
8.2.5	Plant Organization, Operating Instructions, Quality Assurance	122
8.3	Summary Assessment	123
9	**Summary**	**124**
10	**Comment by the USSR Ministry for Atomic Energy Industry on the Safety Assessment of Unit 5 of the Greifswald Nuclear Power Station**	**128**
10.1	Introduction	128
10.2	Comment on Annex A.3 by the Main Designer and the Scientific Leader	129
10.2.1	Comment Regarding Upgrading Measures	129
10.2.2	Comment on Analyses and Verifications	131
10.2.3	Comment on the Documentation	133
10.3	Comment on Annex A.3 by the General Project Engineer	134
10.3.1	Comment on Upgrading Measures	134
10.3.2	Comment on Analyses and Verifications	137
10.4	Conclusions	137

ANNEX

A.1	**Nuclear Power Plants of the type WWER-440/W-213**	**139**
A.2	**Participating Firms and Institutions**	**141**
A.3	**Summary of the Upgrading Measures Derived from the Studies, and of the Analyses and Documents Needed for further Investigations**	**142**
A.3.1	Upgrading Measures	142
A.3.1.1	Materials	142
A.3.1.2	Process Engineering	144
A.3.1.3	Electrical Engineering	150

A.3.1.4	Instrumentation and Control	152
A.3.1.5	Civil Engineering Aspects	158
A.3.1.6	Administration and Operations Management	160
A.3.2	Analyses and Verifications	161
A.3.2.1	Materials	161
A.3.2.2	Process Engineering	164
A.3.2.3	Electrical Engineering	169
A.3.2.4	Instrumentation and Control	170
A.3.2.5	Civil Engineering Aspects	171
A.3.2.6	Administration	171
A.3.3	Documentation and Data	171
A.3.3.1	Materials	171
A.3.3.2	Civil Engineering Aspects	173
A.3.3.3	Administration	174
A.4	**Analyses of Fuel Rod Behaviour by the Kurchatov Institute**	**175**
A.4.1	Calculating the Critical Heat Flux of Fuel Rods in type WWER Reactors	175
A.4.2	Studies of Fuel Rod Behavior in Accidents	177
A.4.2.1	Software Covering Fuel Rod Behavior in Accidents	177
4.2.2.2	Code Verification by Experimental Data	178
A.4.2.3	Calculated Analysis of Fuel Rod Behavior in WWER during Accidents	178
A.4.2.4	Experimental Data on the Physico-Mechanical Properties of Fuel Rod Materials under Normal and Accident Conditions	179
	References, Annex A.4.2	181
A.4.3.	Modeling Fuel Rod Behavior in WWER Reactors under Normal Operating Conditions	183
A.4.3.1	Description of the PIN-Micro Code	183
A.4.3.2	Verification of the PIN-Micro Code	184
A.4.3.3	Optimization of the Initial Gas Pressure in Fuel Rods for WWER-440 Reactors of Increased Power	185
	References, Annex A.4.3	186
	Diagrams, Annex A.4	187

1 Introduction

Comprehensive safety assessments are at present being performed of Unit 5 of the Greifswald Nuclear Power Station by the Gesellschaft für Reaktorsicherheit (GRS) mbH working on behalf of the German Federal Ministry for the Environment, Nature Conservation, and Reactor Safety (BMU). These activities serve to find out to what extent criteria contained in the safety regulations and technical codes applying in the Federal Republic of Germany are met in the engineered safeguards design of the plant.

Pressurized water reactors of the Soviet type WWER-440 were built on the Greifswald site. In the final stage of completion, a total of eight units of this type, each with 440 MW of electric power, had been planned for this site. Units 1-4, commissioned in 1973-1979, are equipped with plants of the older type WWER-440/W-230. They were decommissioned in 1990. Units 5-8 are to be equipped with plants of the advanced type WWER-440/W-213. Construction of these units was begun in the late seventies (Unit 5) and early eighties (Units 6-8) respectively. A commissioning licence for Unit 5 was issued in 1988. Units 6-8 are still under construction, having reached different levels of completion.

Plants of the type WWER-440/W-213 are under construction, or have been in operation for a number of years, in various countries. A list of these nuclear power plants is found in Annex A.1.

The assessment of Unit 5 was begun in the summer of 1990. The investigations have been subdivided into three steps:

(1) Expert assessment of the engineered safeguards design of the plant.

(2) Determination of those engineered safeguards requirements or verifications and upgrading measures which are still required on the basis of existing safety regulations, nuclear codes and standards, and engineered safeguards practice in the Federal Republic of Germany.

(3) More extended safety analysis demonstrating an adequate and balanced engineered safeguards design. In this phase, probabilistic methods are also to be used.

The present report contains results obtained in the first two assessment steps.

The findings refer to Unit 5. As Units 6-8 are largely identical in design to Unit 5, the conclusions and recommendations derived from the investigations in principle can be applied to those units as well.

In the course of these investigations, GRS awarded a number of subcontracts to other institutions, among them various Technical Inspectorates (Technische Überwachungsvereine) and the State Materials Testing Institute (MPA) in Stuttgart. A list of these subcontractors is contained in Annex A.2. The investigations were also supported by Kernkraftwerks- und Anlagenbau AG, Energiewerke Nord AG, and Bauakademie Berlin.

The work is based on close international cooperation with various institutions.

The cooperation with Soviet institutions is of particular importance. It has been organized within the framework of a German-Soviet Government Agreement and additional agreements on cooperation in the field of technical nuclear safety. In the course of the safety assessments of Unit 5 of the Greifswald Nuclear Power Station, a number of German-Soviet joint project discussions were held about the findings made in the course of the investigations and the conclusions derived from them. On the Soviet side, these meetings were attended by representatives of the Ministry for Atomic Energy Industry, the State Supervisory Committee, the Moscow Kurchatov Institute, and Energoprojekt, Moscow, as project engineer and OKB Gidropress as plant designer.

An expert opinion about the investigations carried out on Unit 5 of the Greifswald Nuclear Power Station (WWER-440/W-213) was worked out by the Soviet side. As a result it was found that both sides basically agreed in their assessment of the findings of the investigations and in the conclusions, the recommendations and the backfitting measures to be derived from them. The Soviet comment is published as part of this report (Section 10).

In conclusion of the discussions which were held between the Ministry for Atomic Energy (USSR) and GRS (Federal Republic of Germany) both sides agreed on publishing a joint report on the investigations into the safety of Unit 5 of the Greifswald Nuclear Power Plant in Russian and German language.

The investigations of the WWER reactors are also conducted in close cooperation with the French Institut de Protection et de Sûreté Nucléaire (IPSN), Paris. In the course of the studies on Unit 5 of the Greifswald Nuclear Power Station, joint technical discussions of various subjects were held by GRS and IPSN. In addition, representatives of IPSN took part in German-Soviet project meetings.

The IPSN-conducted studies on Unit 5 of the Greifswald Nuclear Power Station mainly deal with finding out to what extent nuclear power plants of the type WWER-440/W-213 meet the criteria of French nuclear codes and standards. The results of the studies are summarized in a report by IPSN.

The results of the studies carried out by GRS and IPSN were discussed and compared at a joint project meeting in Berlin on March 7 and 8, 1991. Both sides largely agreed in their evaluations of the findings, the recommendations to be derived from them, and the upgrading measures to be taken. The most important conclusions from both studies are to be summarized in a joint report.

In order to clarify technical issues, discussions with the operator of the Paks Nuclear Power Station in Hungary were held in the course of the studies.

To explain the contents of the technical sections (Sec. 4-8), the technical features of installations and systems of nuclear power plants of the type WWER-440/W-213 are described in Section 2. There, the most important engineered safeguards installations are referred to.

In section 3 an overview is given of the present licensing situation of Unit 5 as well as of the most important German codes and regulations in the field of nuclear engineered safeguards.

Sections 4-8 summarize the results of the technical investigations. A more detailed discussion of the investigations is contained in technical working reports.

Section 4 contains an assessment of the core design and the pressurized components. Section 5 deals with accident studies, analyses of the effectiveness of the safety systems, and calculations of the radiological consequences of accidents. Section 6 covers assessments of the safety-related design for in-plant accidents. Section 7 features findings with respect to spreading impacts, civil engineering aspects, and operational radiation protection. Section 8 provides a summary evaluation of the operating experience accumulated in the commissioning phase.

Section 9 presents a summary evaluation of the findings of the investigations. The upgrading measures derived from the investigations are listed, and recommendations are expressed about further investigations. A list of all detailed technical measures derived from the investigations as well as of the analyses and documents still required is contained in Annex A.3.

Section 10 contains the Soviet comment on the investigations performed. The comments presented in Sections 10.2 and 10.3 refer to the list of individual measures, analyses, and documents shown in Annex A.3, where they are listed again in the appropriate places.

2 Technical Plant and Systems Features of the WWER-440/W-213 Nuclear Power Plants

Nuclear power plants equipped with Soviet pressurized water reactors of the type WWER-440/W-213 constitute an advanced version of the type WWER-440/W-230. Compared to the type WWER-440/W-230 nuclear power plants, the facilities of the type W-213 have improved safety features. The most important design features and engineered safeguards of the type WWER-440/W-213 plants are described below.

Figure 2-1 shows a map of the Greifswald Nuclear Power Station with Units 5-8. Like the older W-230 plants (Units 1-4), these units were built as double-unit plants in which two reactors are located in one common reactor hall. One common turbine building serves all the eight units. Figure 2-2 is a cross-section through the building of a type W-213 plant. In contrast to the confinement system with pressure relief valves employed in the W-230 plant, the reactor building of the W-213 plant is connected to a pool-type pressure suppression system (Fig. 2-3).

The type WWER-440/W-213 plants have the following systems for heat removal:

- Primary System:
  The primary system serves for reactor cooling. It is contained inside the confinement. Figure 2-4 shows a design scheme, Fig. 2-5 a diagram of the primary system with the layout of the components.

- Secondary System:
  The secondary system is responsible for transferring the power from the steam generators to the turbines. Most of it is situated in the turbine hall. Figure 2-6 is a design scheme.

- Cooling Water System:
  The circulating water system removes the heat from the condensers. Two service water systems with two component cooling systems serve to cool further operational systems and engineered safeguards. The service water systems are fed by sea water.

Table 2-1 lists the most important design data of the WWER-440/W-230 and W-213 plants. For comparison, the corresponding data of a WWER-1000/W-320 plant and a convoy-1300 plant are also included in Table 2-1.

2.1 Primary System

The primary system (Fig. 2-4 and Fig. 2-5) consists of a water-cooled and water-moderated pressurized water reactor (PWR) with a thermal power of 1375 MW and six main coolant loops (MCL) of the nominal bore 500 (NB 500).

Each main coolant loop is connected to one main coolant pump (MCP), one steam generator (SG), two main gate valves (GV) isolating the reactor from the steam generator. One safety valve (NB 15) is installed in that part of each main coolant loop which can be isolated, in this way providing protection against overpressure.

In order to equalize fluctuations of pressure and volume, the reactor cooling system has a pressurizer (P) connected to the hot leg of a main coolant loop by two connecting pipes of nominal bore 200 (NB 200) in the part which cannot be isolated. The pressurizer spray pipe is connected to the cold leg of the same recirculation pipe by means of an NB 100 pipe. The pressurizer is equipped with two safety valves discharging into a relief tank protected against overpressure by a rupture membrane.

- Reactor Pressure Vessel

Figure 2-7 shows the reactor pressure vessel (RPV). It is an upright cylindrical vessel with a dished head and bottom. The cylinder is made up of three seamless forged rings welded together by circumferential welds. The rings are followed by two nozzle rings and the top flange, which are also connected by circumferential welds.

The pressure vessel is made of low-alloy ferritic steel and is cladded with an austenitic liner. The upper nozzle level of the upper nozzle ring contains six NB 500 outlet nozzles to connect the hot legs, while the bottom nozzle level of the lower nozzle ring is equipped with six NB 500 inlet nozzles to connect the cold legs of the main coolant loops. In addition, both nozzle levels feature two NB 250 nozzles to connect four core flooding tanks and, on the upper nozzle level, one NB 250 nozzle to connect measurement lines.

- Main Coolant Pumps

The main coolant pumps of the W-213 plants differ considerably in their design from the pumps used in the W-230 plants. The main coolant pump is a vertical, single-stage centrifugal pump with a mechanical shaft seal covered with sealing water. The motor bearings and the pump top bearing are lubricated by an oil system connected to the emergency electric power supply. The pump bottom bearing is cooled and lubricated with water from an independent cooling system.

In contrast to the pumps of the W-230 plants, those of the W-213 plants have an additional flywheel mass attached to the motor which, in case of power failure, ensures a sufficiently smooth coastdown of the coolant flow.

- Steam Generators

The steam generators of the W-213 plant, like those of the W-230 plants, are large-volume horizontal vessels with horizontal heat exchanger tubes (heater tubes). Figure 2-8 shows a steam generator.

On the primary side, the coolant enters and leaves through two collectors coming in at the bottom. The steam collector above the steam generator is connected to the steam plenum of the steam generator by five nozzles. Feedwater is supplied to the secondary side of the steam generator through an NB 250 pipe and, separately, by the emergency feedwater supply system with an NB 80 pipe.

Compared to the vertical steam generators used in western pressurized water reactors, the horizontal steam generators have one major advantage in their comparatively large flash area, which results in a lower flash rate and simpler water separation. On the other hand, this design makes it difficult to measure the steam generator filling level, thus exerting a negative impact on feedwater control.

2.2   Secondary System

Figure 2-6 shows a survey diagram of the secondary system, the main steam system and the feedwater system. The pipes of the main steam and feedwater systems are run together out of and into the confinement system respectively, on the 14.7 m platform. The same platform carries all the associated valves (atmospheric main steam dump station, safety valves, quick-closing isolating valves, and feedwater control valves).

- Main Steam System

The plant has six steam generators (SG) and two turbo generators (TG). Each turbo generator is connected to three steam generators and three main steam pipes.

The main steam lines of the two turbo generators are connected to a main steam collector without any isolating valves in between. The main steam collector can be separated into two half systems by a double isolation valve which is open during normal operation.

Each steam generator has two medium-operated safety valves (2 x 50%) with one spring-loaded solenoid/spring control valve each, which are equipped with an electromagnetic additional load. The control valves, and thus also the main valves, can be operated from the control room of the unit.

In the direction of flow, each main steam line contains one quick-closing valve, one check valve, and one isolating valve.

In the section of the main steam line which cannot be isolated, each steam generator has assigned to it one atmospheric main steam dump station (BRU-A) from which the steam can be discharged into the atmosphere (venting over the roof). These dumps are connected to the emergency power system and remove the heat on the secondary side in case a turbine condenser were to fail, especially in the emergency power case.

In case of a turbine trip, the main steam is passed to the turbine condensers through two bypass stations (BRU-K) per half system, provided the turbine condensers are available. The capacity of these bypass stations is roughly 70% of the design capacity of the steam generators.

- Feedwater System

Five feedwater pumps (FP) are installed, which are connected by one collector each on the intake and the outlet sides. The intake and outlet side collectors are split into two half systems separated by isolating valves during normal operation.

There are two sets of two feedwater pumps. Each set is pumping out of its own feedwater tank. The two sets are allocated each to three steam generators. The fifth feedwater pump, which is a standby pump, can pump from either feedwater tank (half systems) when necessary. Per half system, the pumps feed through a pressure pipe downstream of the high-pressure pre-heaters (HPP) into the upper feedwater collector which, like the main steam collector, is split into two half systems not separated during normal operation. From this collector, the feedwater pipes then branch off to the steam generators (Fig. 2-6).

Upstream of the high-pressure pre-heater (HPP) there are two parallel control valves for normal load and low load per half-system leg. The controlled variable is the feedwater flow as a function of the turbine power, and the filling level of the respective feedwater tank, when the filling level limit is underrun. In addition, for filling level control, there is another feedwater control valve in the feeding line to each steam generator on the 14.7 m platform.

- Make-up Feedwater and Condensate System

Make-up feedwater is used to fill up the feedwater system and make up for any losses encountered in operation. It is made available by the central feedwater supply and chemical water treatment systems.

The make-up feedwater enters the low-level condensate tank through control valves. The controlled variable is the filling level in the respective feedwater tank. The condensate is pumped into the condenser from the low-level condensate tank. From the condenser, the condensate is pumped by the condensate pumps into the feedwater tanks through a control valve maintaining the filling level in the turbine condenser.

2.3 Cooling Water Systems

- Circulating Water System

The circulating water system has one circulating water pump per turbine. The pumps take in sea water from the intake and pump building and pump it into the turbine condensers.

- Service Water Systems

There are two service water systems (NKW-A, NKW-C), with A supplying safety-related loads and C supplying plant loads. These A and C systems, inter alia, cool the component cooling systems, NKW-B and NKW-D.

2.4 Engineered Safeguards Design

The safety assessment of Units 1-4 (WWER-440/W-230) of the Greifswald Nuclear Power Station revealed major defects in engineered safeguards design. Despite these deficiencies, the Soviet nuclear power plants of the type WWER-440, irrespective of specific lines, have safety-related characteristics which must be judged positively. Items to be mentioned in this respect are

- the relatively low power density of the reactor core,

- the relatively large water volume in the primary system and on the secondary side of the steam generators, and

- the isolation capability of the main coolant loops.

Compared with the type W-230, the younger type W-213 is equipped with considerately improved engineered safeguards. Thus, their safety systems have higher capacities and, for the most part, are designed as redundant 3 x 100% systems. They are largely separate from the operating systems.

The type W-213 has an emergency core cooling system and a residual heat removal system designed to cover the entire spectrum of possible leakage accidents up to the double-ended break of a main circulation pipe. The plants of the type W-213 have a pressure-resistant compartment system with a pool-type pressure suppression system, the so-called confinement system. Also this system has been designed against the double-ended break of a main circulation pipe.

2.4.1 Emergency Core Cooling and Residual Heat Removal System

The emergency core cooling and residual heat removal system of the primary system serves to feed borated coolant and remove heat especially in loss-of-coolant accidents.

The schematic flow diagram of the emergency core cooling and residual heat removal system and the sprinkler unit of the confinement system are shown in Fig. 2-9.

As a passive core flooding system, four hydraulic accumulators of 40 m^3 of coolant volume each are available. At a pressure of 5.4 MPa, which is generated by a nitrogen blanket, the hydraulic accumulators feed straight into the reactor pressure vessel through separate pipes.

The active systems available are high-pressure (HP) and low-pressure (LP) safety injection systems. In a loss-of-coolant accident, the pressure in the confinement system is suppressed by a sprinkler system. The systems consist of three legs each, which are demeshed. One leg each of the HP, LP and sprinkler systems constitute one unit with common emergency electric power and cooling water supplies.

For each leg of the HP safety injection system there is one 65 m^3 boric acid storage tank (40 g of boron/kg of water); for each leg of the LP safety injection system, a tank holding 500 m^3 of boric acid (12 g of boron/kg of water) is provided.

As soon as one of the 65 m^3 tanks is empty, the HP feed system of the respective leg automatically changes over to the 500 m^3 tank of the LP feed system. The HP safety injection system can function over the entire pressure range (12.2-0.1 MPa). Consequently, there is another feed system available in the LP range (<0.7 MPa) besides the LP pumps.

When the 500 m^3 boric acid storage tank is empty, the LP and the HP safety injection systems change over to the sump recirculation mode. In this mode of operation, the

LP and the HP systems can transfer the residual heat to the service water system for long periods of time.

The pumps of the sprinkler system initially take water from the boric acid storage tanks of the LP system and, after changeover to the sump recirculation mode, from the sump of the building.

2.4.2 Emergency Feedwater System

The emergency feedwater system serves to feed the steam generators in case of failure of the main feedwater system.

A schematic diagram of the system is shown in Fig. 2-10. Starting with the emergency feedwater pumps, the system consists of three legs. Each emergency feedwater pump feeds into two steam generators. One of the three legs is run into the confinement system over the 14.7 m platform, while the other two legs are routed through the rail corridor between Units 5 and 6.

The three pumps of the emergency feedwater system are supplied through a common pipe from a demineralized water tank of 1000 m^3 volume set up in the open air. This demineralized water tank is equipped with a heater preventing the water from reaching temperatures below 5 °C.

All emergency feedwater pumps are set up in the turbine hall. One pump is located in the feedwater pump area at -2.1 m, while the two others are situated at •0 m in the area of the generator.

A separate pump, the so-called operational feedwater pump, is used for plant startup and shutdown. This pump is fed from the feedwater tank, supplies the feedwater collector, but is not supplied with emergency power, and is not activated automatically.

2.4.3 Service Water System and Component Cooling System

The service water system (NKW-A) and the component cooling system (NKW-B) serve to cool important safety-related systems. The pumps of the two systems are located in the joint intake and pump building of Units 5 and 6.

NKW-A is an open system carrying sea water. It is shown in Fig. 2-11. The system has 2 x 150% intake sections equipped with mechanical cleaning systems (bar screen, coarse screen, belt screen). Both intake sections open into a collector from which the three-leg system is supplied. Each leg is equipped with two pumps (2 x 100%) connected to the emergency electric power supply.

NKW-A removes heat from the

- component cooling system (CCS) of the main coolant pumps,
- emergency power Diesel systems,
- motors of the HP emergency coolant pumps,
- LP emergency system heat exchanger, and
- NKW-B.

The coolers of the emergency Diesel power systems are permanently exposed to the cooling water flow. In addition, NKW-A removes the residual heat from the shutdown condensers of the secondary cooldown systems. In case of loss of off-site power, the shutdown condensers are separated from NKW-A.

The component cooling system (NKW-B) is a single-leg system. The pumps and coolers connected to the emergency power system are each designed to 3 x 50%. The pressure in NKW-B is higher than that prevailing in the systems to be cooled. NKW-B is used to cool these components and systems, among others:

- Make-up system of the primary system,
- spent fuel pit,
- component cooling system for the drives of the reactor control rods,
- bearings of the HP emergency coolant pumps.

2.4.4 Confinement System

The confinement system (Fig. 2-2) is a building structure enclosing the primary system in order to prevent releases of radioactive substances in a loss-of-coolant accident. To limit the pressure buildup in a loss-of-coolant accident, the confinement system is equipped with a pool-type pressure suppression system. The system has been designed to accommodate the maximum pressure buildup associated with a 2A-break of the main circulation pipe. It is quite different from the confinement system used in type W-230 plants.

The pool-type pressure suppression system (Fig. 2-3) is a condensation facility consisting of an inlet shaft, 12 pools filled with borated water stacked at right angles to the shaft, and the air traps. It is connected through four large and two small openings with that part of the confinement system which encloses the primary system. The openings are closed with rupture disks, which open the flow path to the shaft at a pressure difference >5000 Pa. In case of an accident, the steam is passed through the shaft into the water of the pools, where it condenses.

If the pressure above the pools rises by more than 500 Pa, the steam-air mixture in that part of the plant flows through double check valves (NB 500) into the air traps where it is retained.

Each pool is fitted with two NB 250 pressure relief dampers which only open towards the shaft. When there is a pressure rise to 0.16 MPa in the shaft, they are locked in the closed condition.

Condensation of the steam in the water of the pools and on structures inside the confinement system causes the pressure in the shaft to decrease below the pressure above the pools. As a result of this pressure reversal, some of the water is forced out of the pools into the shaft. The water that is being discharged is sprayed through perforated metal sheets below the pools. In this way, it contributes to an additional condensation effect. Further pressure suppression in the confinement system, until a negative pressure relative to the outside atmosphere is reached, is achieved mainly by the sprinkler system.

As the pressure of 0.16 MPa will not be reached by small leaks, a pressure equalization between the shaft and the pool compartments is achieved through the unlocked pressure relief dampers. The water reservoir is preserved for major leaks which could follow.

2.4.5 Ventilation Systems

In line with their functions, the ventilation systems are composed of plants

- in the confinement system (CS),
- outside the confinement system in the reactor building,
- for heat removal from electric plant compartments and control rooms,
- for cooling safety-related supply systems (such as battery compartments, rooms for secure AC power supply).

The facilities in the confinement system mainly serve these purposes:

- Maintaining a negative pressure of 150-200 Pa during normal operation as well as a directed air flow.
- Cooling and dehumidifying the air in compartments designed for overpressures, such as compartments of the steam generators, main coolant pumps, and the reactor shaft.
- Removing radioactive contaminants from the air.
- Filtering the air during reloading and repair.

The ventilation system comprises supply and exhaust air as well as recirculation air systems cooled by the component cooling system (NKW-B). The negative pressure and the air cleaning systems are equipped with aerosol filters and iodine filters or carbon adsorber filters.

2.4.6 Electric Power Supply

- Grid Connection

The units of the Greifswald Nuclear Power Station are connected to the integrated power grid through the 220/380 kV open-air switchyard by two 380 kV double lines and three 220 kV double lines. Unit 5 is connected to the 220 kV grid. Each of the two turbo-generators feeds the 220 kV open-air switchyard through an independent unit transformer. The turbo-generators are equipped with generator power circuit breakers.

Each turbo-generator supplies two 6 kV station service systems through one station service transformer. The station service power requirement of the unit can also be met from the 220 kV grid through the standby transformer. Also the standby transformers of the other units can be used by means of cross connections. The power plant has been designed for isolated operation.

- Station Service and Emergency Power Supplies

The station service power supply facility consists of three 6 kV switching systems normally connected with one each of the three emergency A.C. buses by means of coupler circuit breakers. The emergency power supply system is a three-line (3 x 100%), physically separate system.

One line of the emergency power supply system is designed as follows (Fig. 2-12): The main emergency power distribution system is supplied from an emergency power transformer (6 kV/0.4 kV) fed by the 6 kV unit distribution system which is supplied by a Diesel generator when necessary. The main emergency power distribution system is connected to two secure main distributors by means of thyristors. In normal operation, the main DC distributors are supplied by means of two reversible motor generators (RMG) from the secure main distributors. In case of loss of off-site power, a battery ensures continuous electric power supply until the Diesel generator power supply takes over. After the emergency power Diesel generators have been connected, and also in normal operation, the battery is continuously kept charged.

Power is supplied to the control rod drives through separate transformers. Other loads, such as the unit computer and the core monitoring system, are supplied from separate uninterruptible electric feed systems.

2.4.7 Instrumentation and Control

The instrumentation and control systems (I&C systems) comprise all installations for plant monitoring and control as well as the I&C systems important to safety, made up of the reactor control and protection system (SUS) and the control systems for the safety systems (Fig. 2-13 and Fig. 2-14). The control room for the unit and the standby control room have both operational as well as safety-related duties.

The instrumentation and control installations used for plant operation include the measuring and control systems for plant operation, and the special systems needed to check on special process and plant parameters, and the unit computer.

The special systems include the Hindukush core monitoring system, the noise analysis system (NAS), and a leak monitoring system for the detection of external leakages.

The unit control room contains panels and consoles arranged by systems and holding indication instruments and operating controls. To a lesser extent, the indication instruments and operating controls are integrated in equivalent circuit diagrams. The redundant unit computer together with the loggers is installed in a compartment adjacent to the control room.

A standby control room (emergency control room) is used for scramming the unit and monitoring long-term emergency cooling of the reactor in case of failure of the unit's control room. The control room and the standby control room have equal functions. No priority circuit system or changeover system has been installed.

The instrumentation and control system important to safety comprises not only the control and protection (SUS) system, but also the SAOS/GZ system and other control systems. The SAOS/GZ system consists of the instrumentation and control facilities for the emergency cooling system with a sequenced connection of the emergency power loads as well as the control of the emergency cooling and residual heat removal systems and the leaktightness and confinement isolation.

Other control systems serve the atmospheric main steam dump stations (BRU-A), the quick-acting main steam isolating valves, and the additional magnetic loads of the pressurizer safety valves.

The instrumentation and control system important to safety consists of the activation, logic, and control levels. At the activation level, the process variables serving for accident detection are measured, converted into analog electrical signals, and fed to trip units. In principle, there is no multiple signal initialization. The trip units used are motor compensators indicating the current measured value on a scale and mechanically actuating adjustable limit switches.

The activation level is normally supplied with 220 VAC. The logic part contains the circuits for evaluation (e.g., two-out-of-three, one-out-of-two) and coupling of the activation signals and, in the SAOS/GZ area, the sequenced connection of the emergency power loads. It is built in 220 V relay technology. The logic part of the control and protection (SUS) system operates in the closed-circuit mode, while the logic sections of the other control systems employ the open-circuit principle.

At the control level, the drives are controlled individually, normally by coupling safety-related (priority) signals and plant operating signals.

As part of the control and protection (SUS) system, the reactor protection system initiates sequenced actions in cases of malfunction and accident. These actions range between the blocking of power increases, reductions of reactor power by inserting or dropping groups of control rods, and reactor scramming. When a reactor scram is initiated, the drives of the control and shutdown assemblies are de-energized, and the latter fall into the core by gravity force.

The control level is always implemented in 220 V relay technology.

Like the safety systems in the process sections, the instrumentation and control system important to safety is made up of three trains, except for the control and protection (SUS) system. Each train consists of two channels. Within a channel, measured data acquisition and limit signal generation in principle are triple steps with a two-out-of-three selection and, in the control and protection (SUS) system, in special cases even quadruple steps with a two-out-of-four selection.

In the control and protection (SUS) system, a two-out-of-six selection has been provided for the steam generator filling level and the primary system temperature, which is in accordance with the process engineering design (six loops), the measurements per loop being conducted in a one-out-of-one mode. The output signals of the two channels of a train are interconnected in a one-out-of-two mode and initiate the respective protective actions.

The trains of the instrumentation and control system important to safety are located in separate places and are supplied with power by separate systems. The two channels within one train are set up in common compartments and are powered by the same source.

The reactor protection system (part of the control and protection (SUS) system) is made up of two physically separate trains with independent power supplies.

The instrumentation and control system important to safety is designed and made largely independent of the operations instrumentation and control system.

Simultaneous checking in the logic parts of the two channels of one train is blocked by the appropriate circuit. Simultaneous checking in several trains must be prevented administratively. Self-monitoring with respect to safety-related failures is possible only at the activation level in the form of analog signal comparators (transducer mismatch) and monitoring of the measuring channel voltage supply. The limit settings are not monitored.

The power supply of the control systems important to safety comes from main distributors which are supplied uninterruptedly.

2.5 Figures, Section 2:

2-1: Building layout, Nord III/IV Nuclear Power Station (Units 5-8, WWER-440/W-213)

2-2: Sectional view of the building of a WWER-440/W-213 plant

2-3: Detail of the pool-type pressure suppression system, WWER-440/W-213

2-4: Primary system, WWER-440/W-213

2-5: Component arrangement in the primary system, WWER-440/W-213

2-6: Main steam and feedwater system, WWER-440/W-213

2-7: Reactor pressure vessel, WWER-440/W-213

2-8: Steam generator, WWER-440/W-213 (longitudinal section)

2-9: Emergency core cooling and residual heat removal system, WWER-440/W-213

2-10: Emergency feedwater system, WWER-440/W-213

2-11: Service water system (NKW-A) and component cooling system (NKW-B), WWER-440/W-213

2-12: Station service and emergency power supplies, WWER-440/W-213

2-13: Simplified signal flow diagram of the control and protection (SUS) system, WWER-440/W-213

2-14: Signal flow diagram of the control systems (SAOZ/GZ), WWER-400/W-213

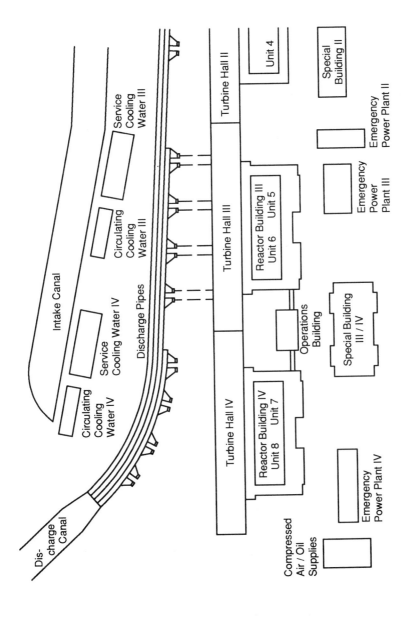

Figure 2-1: Building layout, Nord III/IV Nuclear Power Station (Units 5-8, WWER-440/W-213)

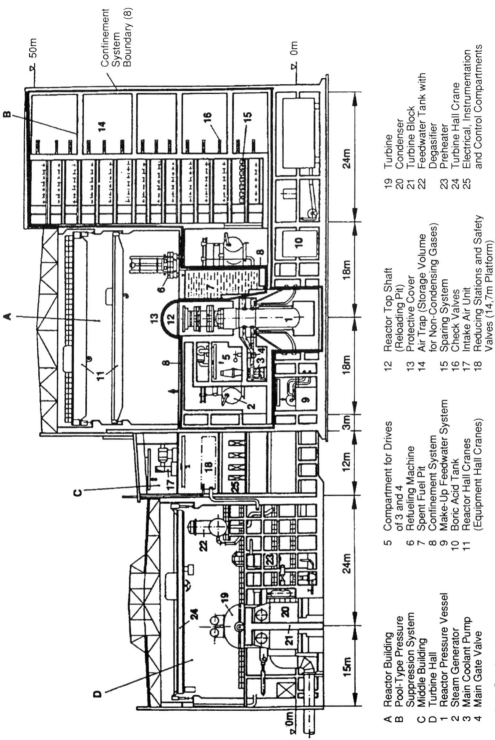

A   Reactor Building
B   Pool-Type Pressure Suppression System
C   Middle Building
D   Turbine Hall
1   Reactor Pressure Vessel
2   Steam Generator
3   Main Coolant Pump
4   Main Gate Valve
5   Compartment for Drives of 3 and 4
6   Refueling Machine
7   Spent Fuel Pit
8   Confinement System
9   Make-Up Feedwater System
10 Boric Acid Tank
11 Reactor Hall Cranes (Equipment Hall Cranes)
12 Reactor Top Shaft (Reloading Pit)
13 Protective Cover
14 Air Trap (Storage Volume for Non-Condensing Gases)
15 Sparing System
16 Check Valves
17 Intake Air Unit
18 Reducing Stations and Safety Valves (14,7m Platform)
19 Turbine
20 Condenser
21 Turbine Block
22 Feedwater Tank with Degasifier
23 Preheater
24 Turbine Hall Crane
25 Electrical, Instrumentation and Control Compartments

Figure 2-2: Sectional view of the building of a WWER-440/W-213 plant

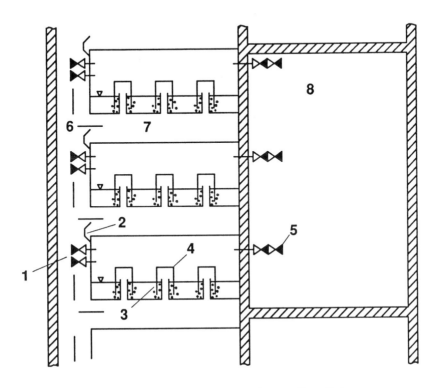

1 Pressure Relief Damper, 2 x NB 250 Parallel
2 Perforated Sheet for Water Sparging
3 Pressure Suppression Pool
4 Deflection Hood
5 Check Valves, 2 x NB 500, in Series
6 Inlet Shaft, Pressure Suppression Pool
7 Inlet Duct to Pressure Suppression Pool
8 Air Trap

Figure 2-3: Detail of the pool-type pressure suppression system, WWER-440/W-213

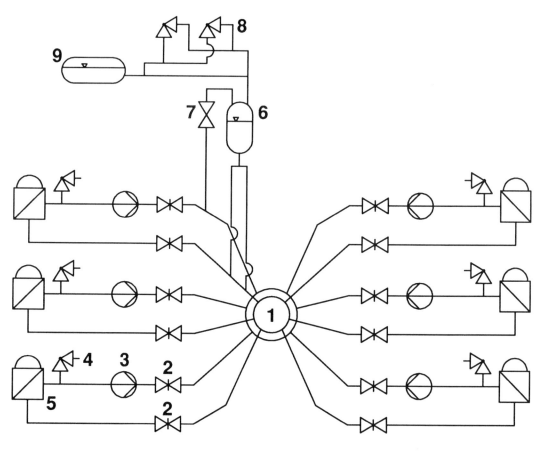

1 Reactor
2 Main Gate Valve
3 Main Coolant Pumps
4 Loop Safety Valves
5 Steam Generator
6 Pressurizer
7 Injection Valve Station
8 Pressurizer Safety Valves
9 Pressurizer Relief Tank

Figure 2-4: Primary system, WWER-440/W-213

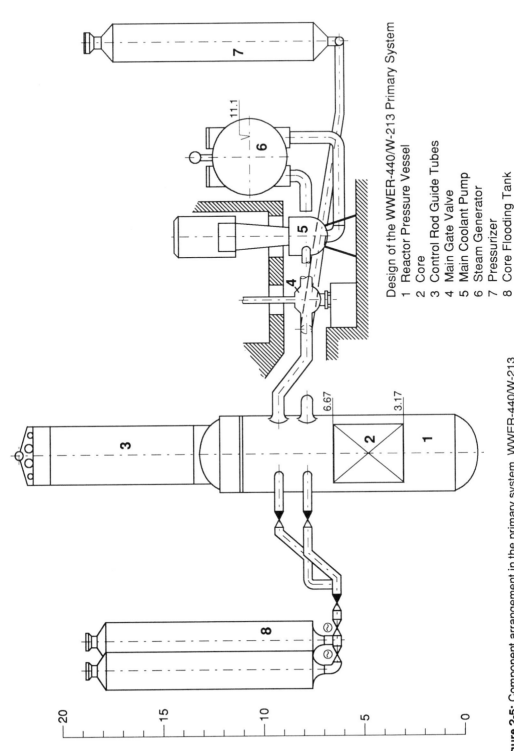

Figure 2-5: Component arrangement in the primary system, WWER-440/W-213

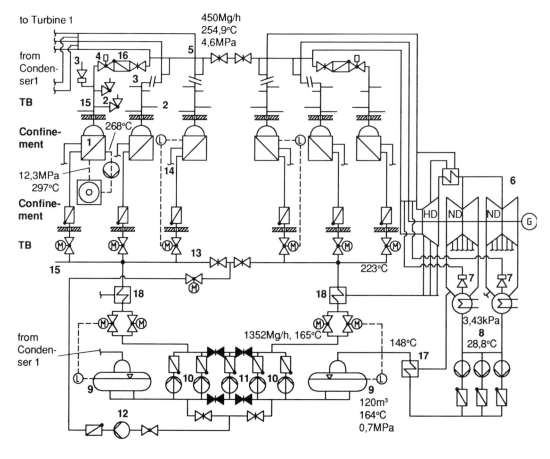

1. Steam Generator
2. Steam Generator Safety Valves
3. Steam Dump Station (Atmosphere)
4. Quik-Action Isolating Valves
5. Common Steam Collector Feed
6. Turbines, Generators
7. Steam Dump Stations (Condenser)
8. Condensers / Condensate Pumps
9. Feedwater Tank Heater
10. Feedwater Pumps
11. Feedwater Standby Pump
12. Feedwater Operating Pump
    (Startup, Shutdown)
13. Upper Feedwater Pressure Heater
14. from Emergency Feed System
15. to Cooldown System
16. Check Valve
17. Low-Pressure Preheater
18. High-Pressure Preheater

Figure 2-6: Main steam and feedwater system, WWER-440/W-213

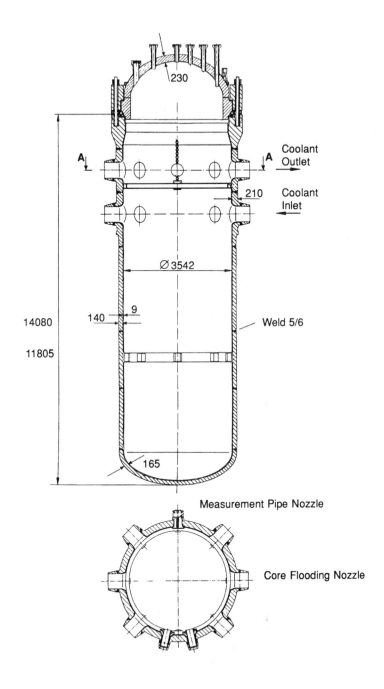

Figure 2-7: Reactor pressure vessel, WWER-440/W-213

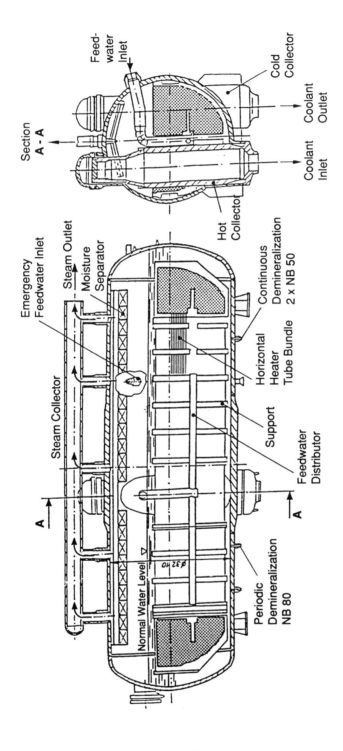

Figure 2-8: Steam generator, WWER-440/W-213 (longitudinal section)

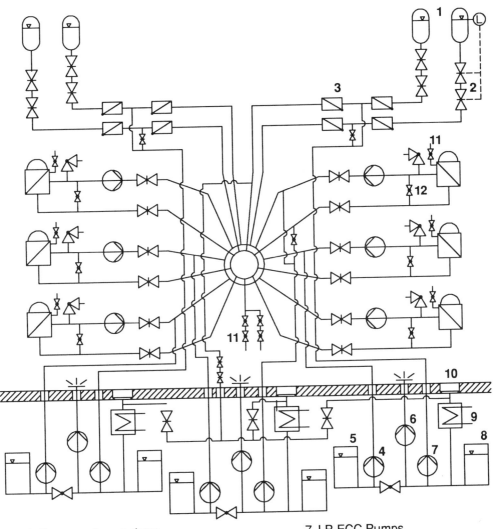

1 Pressure Accumulator
2 Scram Isolating Valves
3 Check Valves
4 HP-ECC Pumps
5 Boric Acid Storage 65 m^3
6 Sprinkler Pumps
7 LP-ECC Pumps
8 Boric Acid Storage Tank 500m^3
9 Emergency Cooler
10 Building Sump
11 Accident Degasing
12 Accident Drainage Tank

Note: All isolating valves shown here are motor-driven

Figure 2-9: Emergency core cooling and residual heat removal system, WWER-440/W-213

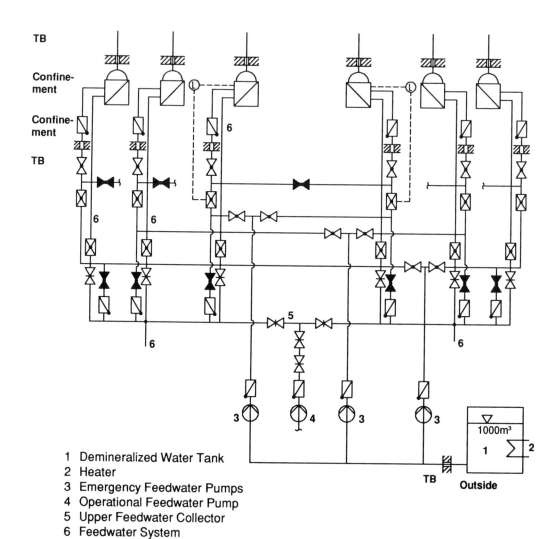

1 Demineralized Water Tank
2 Heater
3 Emergency Feedwater Pumps
4 Operational Feedwater Pump
5 Upper Feedwater Collector
6 Feedwater System

Figure 2-10: Emergency feedwater system, WWER-440/W-213

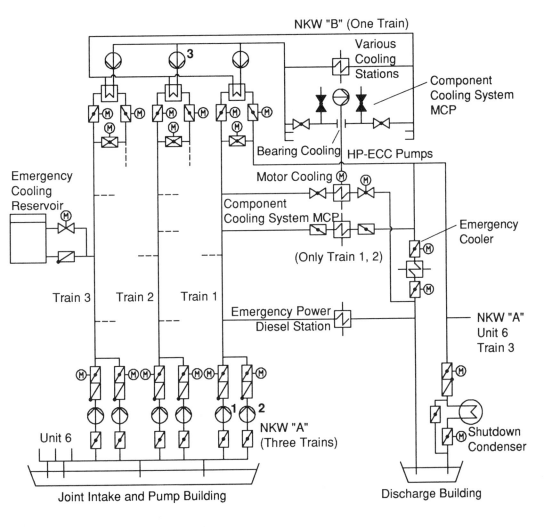

1 NKW "A"-Pump (3 • 100%)
2 NKW "A"-Standby Pump (3 • 100%)
3 NKW "B"-Pumps (3 • 50%)
   (Feeding One Train)

Figure 2-11: Service water system (NKW-A) and component cooling System (NKW-B), WWER-440/W-21.

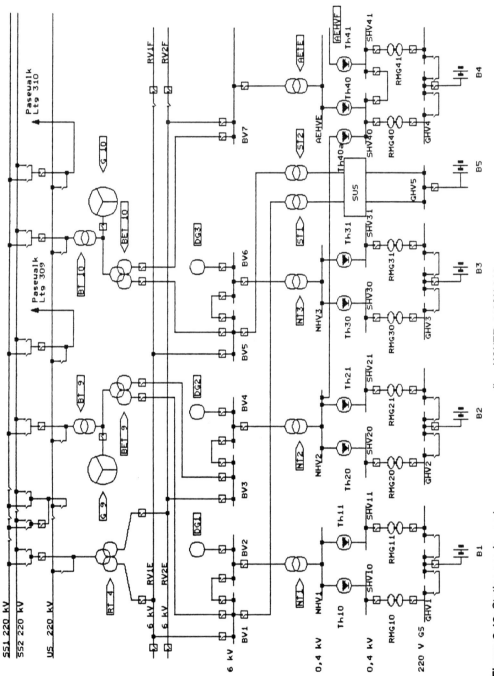

Figure 2-12: Station service and emergency power supplies, WWER-440/W-213

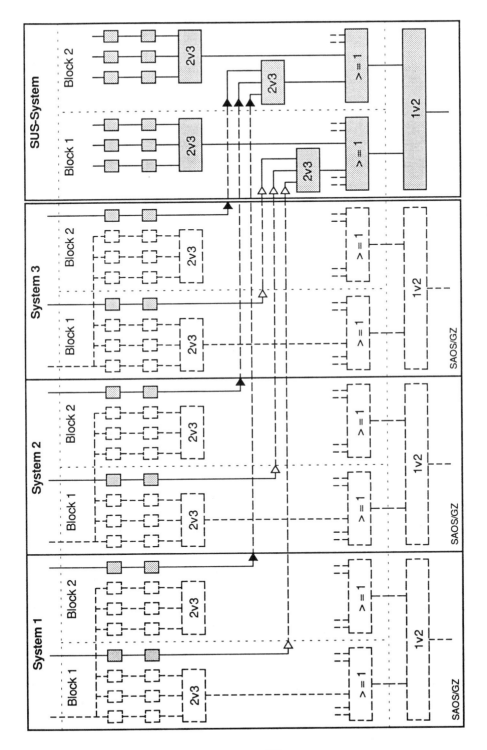

Figure 2-13: Simplified signal flow diagram of the control and protection (SUS) system, WWER-440/W-213

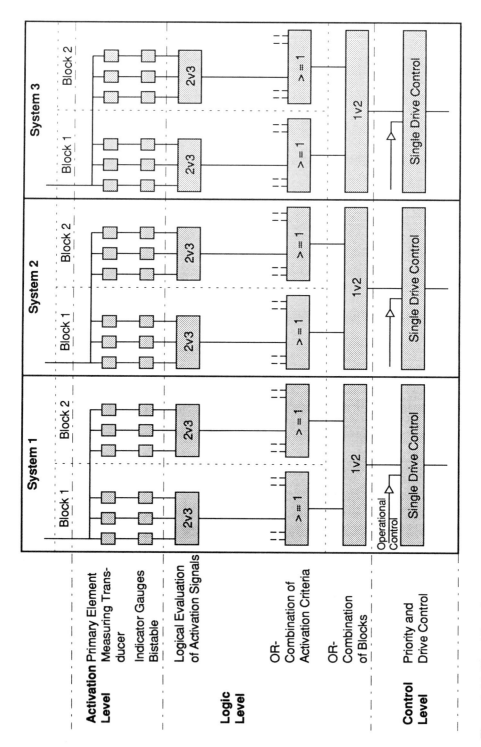

Figure 2-14: Signal flow diagram of the control systems (SAOZ/GZ), WWER-440/W-213

Table 2-1: Comparison of Important Design data of WWER and Convoy Plants

Feature	Dim.	WWER-440 W-230/W-213	WWER-1000 W-230	Convoi-1300 KKE
Thermal power	MW	1 375	3 000	3 765
Gross electric power	MW	440	970	1 314
Net efficiency	%	30	30	33
Reactor core				
- Core average heat flux	W/cm²	43.8	56.8	61.2
- Mean power density	kW/l	86	107	93
- Mean burnup	MWd/kg	29	40	45
Fuel assemblies				
- Type	-	hexagonal with shroud	hexagonal without shroud	square without shroud
- Number	-	312	163	193
Fuel rods				
- Numer/fuel assembly	-	126	312	236
- Active length	mm	2 500	3 550	3 900
- Outside diameter	mm	9.1	9.1	10.75
- Wall thickness	mm	0.65	0.65	0.73
- Cladding tube material	-	Zr + 1% Nb	Zr + 1% Nb	Zircaloy-4
Control assemblies				
- Number	-	37	61	61
- Type	-	Double assemblies	18 control rods per assembly	24 control rods per assembly
- Absorber material	-	Boron steel	B_4C	Ag80In15Cd5
Primary system pressure	MPa	12.2	15.7	15.7
Secondary system pressure	MPa	4.6	6.3	6.4
Coolant temperature: Reactor inlet and outlet	°C	265/295	290/320	291/326

Table 2-1: Comparison of important design data of WWER and Convoy Plants (continued)

Feature	Dim.	WWER-440 W-230/W-213	WWER-1000 W-230	Convoi-1300 KKE
Number of loops	-	6	4	4
Number of turbines	-	2	1	1
Turbine speed	revs./min.	3 000	3 000	1 500
Water volume:				
- In primary system	m^3	215	298	372
with reference to power	m^3/GW	156	99	99
- In secondary system	m^3	252	264	231
with reference to power	m^3/GW	183	88	61
Main coolant pipe				
- Nominal bore	mm	500	850	750
- Material	-	Austenitic steel	Ferrite with austenitic steel cladding	
- Main gate valve	-	existing	not exsisting	
Steam generator type	-	horizontal	horizontal	vertical
- Heater tube diameter	mm	16	13	19.7
- Heater tube metarial	-	Austenitic steel	Austenitic steel	Incoloy
Reactor pressure vessel				
- Diameter	m	4.27	4.51	5.62
- Wall thickness	mm	140	190	250
- Height	m	11.8	10.88 (without lid)	12.4
- Material	-	low alloy steel		low alloy fine-grained steel
		Austenitic steel cladding		

3 Basic Legal Licensing Principles

3.1 Licensing Situation, Unit 5

The GDR State Office for Atomic Safety and Radiation Protection (SAAS) issued these licence:

- In 1977, a licence to construct the plant.

- On December 30, 1988, a licence to commission the unit for trial operation.

After the commissioning licence had been granted, the unit had been in trial operation for 112 days in 1989 at a power not exceeding 55% of the design power; in that phase, it had been synchronized with the grid for some 66 days.

On November 24, 1989, an event occurred in a planned experiment in which the first reactor protection signal for reactor scram was not generated. The SAAS then ordered the commissioning phase to be interrupted pending express clearance.

The subsequent functional tests in the subcritical state revealed other defects in engineered safeguards. For this reason, a continuation of the trial phase of operation has not been permitted so far.

In October 1990, the Joint Institution for Reactor Safety and Radiation Protection of the Federal States (GEL), at that time the competent licensing authority under the Atomic Energy Act, informed the holder of the licence, Kraftwerks- und Anlagenbau AG (KAB) Berlin, that a major prerequisite of continued trial operation was the implementation of the upgrading measures to be derived from the present assessment of Unit 5.

From the date of the first Treaty on the Establishment of a Monetary, Economic and Social Union between the Federal Republic of Germany and the German Democratic Republic, which entered into force on July 1, 1990, the existing licences have a guaranteed legal continuity for five years. New permits must be applied for fully under Sec. 7 of the German Atomic Energy Act.

3.2 Valid Legal Principles of Licensing

The legal framework for the peaceful use of nuclear energy in the Federal Republic of Germany is established by the German Atomic Energy Act. This Atomic Energy Act was adopted in 1959 and has since been amended repeatedly /1/.

Sec. 7, Subsec. 2 of the Atomic Energy Act lists the licensing prerequisites. It says that a licence may be issued only if

- the provisions necessary in the light of the state of the art have been made against damage arising from the construction and operation of the plant,

- the necessary protection against disturbances or other impacts created by third parties is ensured,

- mostly public interests, especially those pertaining to clean water, air and soil, do not stand in the way of the choice of site.

The safety-related licensing prerequisites are not defined in more detail in the Act, but are spelt out in subsequent legal ordinances, technical codes and regulations. The most important are these:

- The Radiation Protection Ordinance /2/

The Radiation Protection Ordinance contains the basic principles of radiation protection. The supreme principle is the requirement to minimize radiation. It implies that radiation exposure and contamination be minimized, in accordance with the state of the art and also in consideration of the conditions prevailing in a specific case, even below defined dose limits. This principle applies both to normal operation and to accident conditions.

- The Safety Criteria /3/

The Safety Criteria for Nuclear Power Plants encompass the basic safety goals to be guaranteed in the design of the plant. In particular, also under unnormal operating conditions, these basic safety requirements must be met: It must be possible to shut the plant down at any time, remove the afterheat over long periods of time, and confine safely all radioactive substances.

- The Accident Guidelines /4/

The Accident Guidelines were set up for more recent nuclear power plants equipped with pressurized water reactors. They apply to plants whose first partial construction licences were issued not before July 1, 1982. Consequently, these Guidelines cannot be referred to directly, but only indirectly, in assessing Unit 5.

On the basis of experience accumulated in safety analysis, expert assessment, and operation of nuclear power plants, the Accident Guidelines define those accidents on which the safety-related design of nuclear power plants with pressurized water reactors must be based, and the verification which must be produced especially with respect to the observance of accident planning levels (dose limits) as specified in Sec. 28, Subsec. 3 of the Radiation Protection Ordinance. Accidents against which provisions must be made by building structures or other means of technical protection are defined independent of the technical design of a plant. The necessary provisions must be made in accordance with the state of the art.

- The RSK Guidelines for Pressurized Water Reactors /5/

On the basis of the fundamental safety goals contained in the Safety Criteria, the German Advisory Committee on Reactor Safety (RSK) has formulated in more detailed and precise guidelines the safety-related requirements to be met in the construction and operation of pressurized water reactors.

These ordinances and guidelines define the requirements and approaches which have proved to work satisfactorily for many years of safety assessment and safety practice of nuclear power plants. The provisions are based mainly on the concepts and designs of light water reactor (especially pressurized water reactor) designs customary in West Germany. However, this does not exclude alternative technical solutions able to meet safety goals, to which the implementing regulations in these codes and standards can be extrapolated analogously. This aspect must be borne in mind when evaluating reactors of different designs, in this case, the plant concept of Unit 5 of the Greifswald Nuclear Power Station.

It must be examined, therefore, whether the existing design satisfies the protection goals underlying the codes, especially whether sufficient provisions have been made to avoid and manage accidents.

Where applicable codes and regulations are not met, it must be investigated whether such deviations give rise to a deficit in safety and, if so, what measures can be taken to make up for such deficiency.

Factors of particular importance in the analyses and safety-related evaluation of the plant are the requirements listed in greater detail in the RSK Guidelines for Pressurized Water Reactors /5/, as for example:

- For the pressurized water reactors designed in West Germany, the maximum accident pressure acting on the containment is determined on the basis not only of the energy and coolant inventories of the primary system, but also of the energy and mass contents of the secondary side of the steam generator.

- Building structures, systems, and components important to engineered safeguards must be designed against external impacts (earthquakes, airplane crashes, etc.).

- The design of, and the requirements to be met by, the reactor scram system (criteria for activation, dropping times of the shutdown rods, design details).

The German technical codes and standards contain design requirements which safety systems must meet in terms of redundancy, diversity, demeshing, and physical separation of the different trains of systems. Above and beyond the criteria contained in the Soviet codes and standards pertaining to safety systems (PBJA-04-74 /6/ and OPB-82 /7/), not only single failures, but also the absence of one level of redundancy because of repair must be included. In addition to the single failure, the Soviet codes and standards assume another hidden failure on components which are not subjected to functional test. This failure will be discovered only during an accident, which may have a negative impact on the sequence of accident events.

In the Soviet codes and standards, the single-failure criterion is restricted to active components. In the German codes and standards, passive components are also taken into account. However, under certain preconditions, the application of the single-failure criterion to passive components can be waived if special requirements are met in terms of reliable design, manufacturing, and surveillance.

In addition to monitoring regulations, the Radiation Protection Ordinance also contains radiation protection provisions, referring e.g. to:

- Radiation protection principles, especially in Sec. 28,

- The protection of the public and of the environment from the hazards of ionizing radiation, especially in Sec. 45,

- Occupational radiation exposure, especially in Sec. 49.

The plant must be examined for compliance with these provisions. In order to be able to assess whether the accident planning levels under Sec. 28, Subsec. 3 of the Radiation Protection Ordinance are met, calculations must be performed of potential radiological accident consequences in accordance with the Accident Guidelines for a number of radiologically relevant design basis accidents. For plants which come under the Accident Guidelines, these radiologically representative accidents must be examined:

- 2A-break in a primary circulation pipe.

- Leakage outside the containment of a measurement pipe carrying primary coolant.

- Leakage with sealing capability in a main steam line outside the containment, accompanied by the simultaneous occurrence of defects in steam generator heater tubes.

- Long-term failure of the main heat sink due to plant leakages in steam generator heater tubes.

- Leakage in a pipe in the offgas system.

- Fuel assembly damage during handling.

- Leakage of a vessel filled with radioactively contaminated water.

- Leakage of a vessel under seismic impacts.

It must be investigated to what extent the list of these accidents can be applied analogously to Unit 5 of the Greifswald Nuclear Power Station.

Accident calculation bases are indicated for analyses of these radiologically representative accidents. Furthermore, the applicability to Unit 5 of the Greifswald Nuclear Power Station of these rules of calculation must be examined.

References, Sec. 3

/1/ Gesetz über die friedliche Verwendung der Kernergie und den Schutz gegen ihre Gefahren (Act on the Peaceful Use of Nuclear Energy and the Protection against its Hazards), (Atomgesetz) (Atomic Energy Act) as promulgated on July 15, 1985 (Bundesgesetzblatt I, p. 1565), last amended as published in Bundesgesetzblatt I, No. 61 on November 10, 1990, p. 2428.

/2/ Verordnung über den Schutz vor Schäden durch ionisierende Strahlen (Ordinance on Protection against Injuries by Ionizing Radiation) (Strahlenschutzverordnung) (Radiation Protection Ordinance) as promulgated on June 30, 1989 (Bundesgesetzblatt I, p. 1321), last amended as published in Bundesgesetzblatt II, No. 35 on September 28, 1990, p. 885.

/3/ Sicherheitskriterien für Kernkraftwerke (Safety Criteria for Nuclear Power Plants), October 21, 1977 (Bundesanzeiger No. 206, November 3, 1977).

/4/ Leitlinien zur Beurteilung der Auslegung von Kernkraftwerken mit Druckwasserreaktoren gegen Störfälle im Sinne des § 28 Abs. 3 Strahlenschutzverordnung (Guidelines for Assessing the Design against Accidents of Nuclear Power Plants with Pressurized Water Reactors in Accordance with Sec. 28, Subsec. 3 of the Radiation Protection Ordinance) (Störfalleitlinien) (Accident Guidelines), October 18, 1983 (Bundesanzeiger No. 245a, December 31, 1983).

/5/ RSK-Leitlinien für Druckwasserreaktoren (RSK Guidelines for Pressurized Water Reactors), 3rd edition, October 14, 1981, including amendments as published in the Bundesanzeiger No. 106 on June 10, 1983, and Bundesanzeiger No. 104 on June 5, 1984

/6/ Nuclear Safety Regulations for Nuclear Electric Power Plants (AES), PBJA-04-74.

/7/ General Safety Regulations of Nuclear Power Plants during Design, Construction, and Operation, OPB-82.

4 Reactor Core and Pressurized Components

4.1 Core Design

The safety requirements pertaining to core design have been laid down in several KTA Codes, especially

- KTA Code 3101:
Designing the reactor cores of pressurized water and boiling water reactors.

Part 1: Basic thermohydraulic design principles.

Part 2: Neutron physics criteria to be met in design and operation of the reactor core and adjacent systems.

- KTA Code 3103:
Criteria to be met by shutdown systems of light water reactors

The requirements include

- two independent shutdown systems to terminate the fission reaction with a sufficient shutdown margin;

- compliance with the safety-related limits so as to prevent failure of the fuel rods;

- the installation of suitable monitoring devices to control reactor core conditions and initiate reactor protection measures.

- KTA Code 3204:
Loads acting upon the internals of the reactor pressure vessel as a result of operational and accidental loads must not jeopardize the safe shutdown capability of the reactor and sufficient cooling of the reactor components.

4.1.1 Neutron Physics Core Design

The core of the reactor is made up of 349 hexagonal fuel assemblies of which 37 assemblies are at the same time control assemblies. The control assemblies comprise a fuel section at the bottom and an absorber section at the top.

The control and protection system (SUS) of the reactor is available for reactor scrams and power limitations. The measures planned for protection of the reactor core are subdivided into four steps (HS-1 to HS-4).

Reactor scram is initiated only by criteria of the first step (HS-1), i.e. all control assemblies are dropped. The activation criteria for steps HS-2 to HS-4 result in reactor shutdown or reactor power limitations.

The neutron physics activations of the reactor protection system are derived from the signals of the out-of-pile neutron flux instruments. Other measurements performed by reactor instruments, such as those of the coolant outlet temperatures of 210 fuel assemblies and in-pile neutron flux measurements at 252 core positions, provide information about the status of the core. They do not initiate any automatic measures.

As a second reactor shutdown system, there is the only non-redundant operational boron injection system (feed system). High-pressure boron injection of the emergency cooling system is initiated only when emergency cooling is required. Activating this system manually requires additional measures to be taken (manual activation of an interlock).

- Defects Recognized and Upgrading Measures Required

(1) The accident analyses confirm that the rate of control assembly insertion is enough when the first activation criterion is effective. It remains to be investigated whether the reactor is shut down reliably in all design basis accidents, also if only the second activation criterion is effective, i.e. the first activation is assumed to fail.

(2) Sticking of the most effective control rod has been taken into account in the design of the reactor scram system. In the Accident Guidelines, the basic assumption is made that operating transients in an assumed failure or partial failure of the reactor scram system are sufficiently unlikely. It remains to be investigated whether the reactor scram system is sufficiently reliable.

(3) Establishment of a redundant, emergency-powered, efficient boron injection system with an injection pressure above the operating pressure.

4.1.2 Thermohydraulic Core Design

The thermohydraulic design must ensure sufficient cooling of the fuel rods. As a parameter characterizing sufficient cooling of the fuel rods, PWRs use the DNB ratio which is calculated for each fuel rod section from the ratio of the critical heat flux density to the actual heat flux density.

The hot channel model is used to take into account the most adverse cooling conditions. The power factors for the hot channel model are determined in reactor physics design calculations.

The relation used to determine the critical heat flux density must be verified experimentally.

Annex A.4, as a supplement to the Soviet comment about this report, contains findings by the Kurchatov Institute, Moscow, about calculations of the critical heat flux and the fuel rod behavior in normal operation and under accident conditions.

- Defects Recognized and Upgrading Measures Required

(1) For the thermohydraulic design, the subchannel factors for the increase in enthalpy, $K_{\Delta H}$, and the heat flux density, K_q, for the current fuel assembly design with exchange cross bores must be substantiated.

(2) The observance of the minimum permissible DNB values must be verified for the design basis transients with regard to the effective power limitation achieved in each case (HS-4 or HS-3). This also requires detailed information about the accuracy of the DNB correlations used.

(3) Power density characteristics must be included in the automatic power limitation and the reactor scram respectively. The algorithms used for power density monitoring must be verified.

4.1.3 Mechanical Design of Reactor Pressure Vessel Internals and of the Core

Verifications in the following areas are required to show compliance with the criteria to be met by reactor pressure vessel internals under KTA 3204:

- Accommodation of the weight and deformation forces of fuel assemblies.

- Ensuring the position and alignment of fuel assemblies.

- Accommodation of the shocks produced by the control assemblies in cases of reactor scram; coolant flow configuration in the reactor pressure vessel.

- Accommodation of the in-pile irradiation specimens for brittle fracture control of the reactor pressure vessel material.

- Ensuring the stability of the core geometry under accident conditions.

Under KTA 3103, the effects of gamma and neutron irradiations on these points must also be considered.

The loads arising in normal operation and under accident conditions were calculated by the Soviet plant vendor, but no documents are available for reconstruction. The loads associated with maximum loss-of-coolant accident were determined as upper limits of potential loads arising from accidents.

According to information provided by Kraftwerks- und Anlagenbau AG (KAB), analytical results have shown the internals of the reactor pressure vessel to withstand the loads arising during normal operation and at maximum loss of coolant.

It is not known whether the embrittlement of materials caused by neutron irradiation in the operation of the reactor was taken into account sufficiently in the calculations. Calculations of load cases arising from external impacts (airplane crash, safe shutdown earthquake, explosion blast wave) have not been carried out. No information is available to show to what extent the possible loads associated with external impacts are described in a sufficiently conservative way by the results determined for a loss-of-coolant accident.

For comparable materials, the internals of the pressure vessel of WWER-440 reactors have roughly the same wall thicknesses as those of KWU pressurized water reactors.

Because of the smaller diameters of WWER reactors, it may be assumed therefore that potential loads arising from accidents in general will be lower than those in KWU pressurized water reactors.

The fuel assemblies of WWER-440 reactors differ from those of KWU pressurized water reactors in their hexagonal (instead of square) cross section, enclosure of the fuel assembly in a shroud, and in the design of the fuel pellets (central bore, longer pellets).

The control assemblies differ from the control rods used in KWU pressurized water reactors in that the control assemblies are not inserted into the permanently positioned fuel assembly, but the fuel section is replaced successively by the absorber section of the control assemblies. The absorber material used is boron steel instead of a silver-indium-cadmium alloy.

Judging from the operating experience accumulated by Units 1-4, the design of the fuel assemblies and control assemblies has proved to be satisfactory in normal operation. Possible influences on the insertion time resulting from the measured slant of the RPV must be verified and their effects on admissibility evaluated.

The materials, other than those used for the cladding tubes, the box, and absorbers, are comparable to those employed by KWU. However, on the basis of past operating experience, also those materials which are not comparable can be considered to be proven in operation.

In accordance with the Soviet standard OPB-82 the following limits must be observed for fuel rod failures:

- For normal operation:

  - Percentage fraction of fuel rods with gas leaks <1% of the total.

  - Percentage fraction of fuel rods with direct contact between the coolant and the fuel <0.1% of the total.

The permissible activity level of the coolant in the primary system will be derived from those limits.

- For greatest possible design basis accidents:

    - Temperature of the fuel rod claddings <1200 °C.

    - Ratio of local depth of oxidation of the fuel rod cladding relative to the cladding thickness before oxidation <18%.

    - Fraction of the mass of the reacting zirconium <1% of the total zirconium mass in the core.

The limits indicated for fuel rod failures during normal operation do not correspond to the requirements specified in German nuclear codes and standards. According to German assessment criteria, the fuel rods must be so designed as to withstand the loads throughout the entire in-pile time, with the planned mode of operation taken into account. The design data available do not indicate that these requirements might not be met by the WWER fuel assemblies. However, no verification of these requirements is available.

The Soviet limits for design basis accidents more or less correspond to the limits included in German codes and standards. In addition to the limits mentioned above, the German evaluation criteria contain two additional requirements according to which the maximum permissible number of failed fuel rods must be limited and cladding tube expansions must be taken into account in verifying sufficient post-decay heat removal. No information is available at present to indicate whether these requirements also were taken into account in the design of the fuel assemblies.

- Defects Recognized and Upgrading Measures Required

(1) The calculations performed by the plant vendor of loads acting on the RPV internals during normal operation and under loss-of-coolant accident conditions must be presented.

(2) The activity level of the coolant in the primary system must be determined from the Soviet limits indicated for fuel rod failures.

(3) Possible influence on the control rod insertion time resulting from the slant of the reactor pressure vessel must be checked again (the slant is about 1.5 mm).

4.1.4 Fuel Assembly Damage during Handling

Due to the fact that the designs are comparable, a boiling water reactor (BWR) fuel assembly (Krümmel Nuclear Power Station) is used as a basis for evaluating handling accidents. Table 4-1 below shows the dimensions and weights involved.

Tabelle 4-1:

	BWR	NPP Greifswald Unit 5
Fuel assembly weight	ca. 300 kg	ca. 215 kg
Fuel assembly length	ca. 4.5 m	ca. 3.2 m
Bos wall thickness	2.5 - 3.0 mm	2.1 mm
Fuel rod fastening design load accommodation	similar with both types, over spacer, support rods and fuel assembly box	
Maximum drop height	ca. 15 m	ca. 7 m

Due to the lower weight, the shorter fuel assembly length, and the lower maximum drop height, the loads to which a fuel assembly in the Greifswald Nuclear Power Station will be subjected in a fuel assembly crash will be lower. Load accommodation is also comparable as a consequence of comparable box wall thicknesses and fuel rod fastening designs.

As a basis for accident calculation in the Accident Guidelines, all fuel rods of an outer edge of the square fuel assembly are assumed to be damaged by incidents and by accidents during fuel assembly handling and storage.

Under the assumption of two outer edges of the hexagonal WWER fuel assembly being damaged, accident assessment must be based on 13 defective fuel rods. As WWER fuel assemblies are handled only with their channels, this is a conservative assumption. Due to its rigidity, the channel is able to accommodate some of the loads arising in a handling accident.

4.2 Pressurized Components in the Primary and Secondary Systems

4.2.1 Purposes

The pressurized equipment items (tanks, pump housings, isolating valve housings) and pipes can be licensed if their integrity during normal operation, operational transients, and accidents can be demonstrated with the necessary safety margins. For this purpose, the following items need to be analyzed:

- Suitability of the materials used.

- Mechanical and thermal loads assumed in structural analyses, including the effects of neutron exposure.

- Technical design details with respect to stress concentrations and inspectability.

- Measures of quality assurance in fabrication, pre-assembly, and assembly.

- Program and methods of as-delivered testing and concept of in-service inspections.

- Interactions of structural materials with plant media.

The following scope of the plant was included in the evaluation:

- Primary systems equipment and pipes held under operating pressure, i.e., reactor pressure vessel, pressurizer, housings of valves and pumps, steam generators, main circulation pipe, pressurization system, and

- equipment and pipes for cooling the nuclear fuel, i.e., emergency core cooling system and residual heat removal system, feedwater and main steam systems of the secondary system.

Pipes of nominal bore NB 250 were evaluated only in specific cases, as modifications or replacement are possible without any restrictions when required.

4.2.2 Safety-related Assessment and Measures Required

Evaluations of technical safety of the pressurized components of the primary and secondary systems were carried out in two steps:
1. Comparison of the criteria set forth in codes and standards.
2. Evaluation of the components in the condition existing in the plant.

- Comparison of the Criteria Set forth in Codes and Standards

The comparison of the criteria contained in codes and standards about the way in which provisions are to be made against failures and defects shows the measures planned to meet these criteria to be more or less sufficient. Compared to German codes and standards, the Soviet codes and standards require

- less depth of verification of satisfactory toughness of the base materials and welds (no requirement to verify the reduction of area in the through thickness direction, no requirement to test the heat-affected zone);

- less depth of verification of non-destructive tests, especially in ultrasonic inspection (e.g. fewer directions of ultrasonic radiation and inspection angles);

- less depth of verification of the resistance of the structure to accommodate the loads actually occurring and those postulated;

- no unlimited inspectability in the area of the weld roots and nozzle structures;

- no direct limitation of the neutron fluence in the area of the reactor pressure vessel wall close to the core. (The neutron fluence is limited indirectly by the brittle fracture transition temperature.)

Each individual component was examined to see to what extent the limitations referred to above can be removed to a sufficient degree by modifications to components and systems, by additional inspection measures and materials examinations, and by steps to reduce influences of operation.

To assess the licensability of the components already completed, it was also examined whether deviations from the criteria set forth in codes and standards existed on such a scale as to render provisions against defects and failures inexistent. In this respect, it must also be borne in mind that technical codes and standards basically are designed so that deviations are admissible in meeting individual criteria, but that such deviations must always be based on identifiable technical reasons about admissibility.

With respect to the compliance with the requirements of the Soviet codes and standards to the components of the primary and secondary systems, no final statement can as yet be made. The documentation existing at the manufacturers' of the components has not yet been inspected; the documents sampled on the construction site were found to be not detailed enough to allow the evaluation of the documented test result. No sufficient information is as yet available about findings elaborated in research programs, e.g. about the reliability of workmanship and the long-term stability of the materials used.

The large components in the primary and secondary systems of Units 5 to 8 largely correspond to the components of the WWER-440/W-230 line in terms of materials and design. In evaluating Unit 5, however, it was only possible to take into account the operating experience accumulated by Units 1 to 4 of the Greifswald Nuclear Power Station.

- Component Evaluation
- Reactor Pressure Vessel

The selected materials and the design of the reactor pressure vessel largely correspond to codes and standards. To limit influences of neutron exposure of the material, measures must be taken to ensure safety margins sufficient also in the long term. With respect to the nozzles to be welded into the head of the reactor pressure vessel, the inspection techniques currently available need to be adapted in a specific way in order to ensure sufficient provisions against damage. Although inspectability is restricted at some points as a consequence of the design, sufficiently representative in-service inspections by non-destructive techniques are possible.

At some points of the reactor pressure vessel of Unit 5, lower toughness values were found which can be assessed only after all documents held by the vendor have been inspected. If necessary, subsequent materials tests will have to be carried out. No limitations are expected to arise from the detailed stress analysis yet to be conducted. The consequences to be drawn from the measured slant of the reactor pressure vessel as far as the loads acting on nozzle connections are concerned cannot yet be assessed.

- Main Coolant Line, Housings of the Main Coolant Pumps and Isolating Valves

In contrast to PWRs of KWU, the main coolant lines in this plant are made of austenitic material. Consequently, the interfaces connecting ferritic and austenitic materials exist in the reactor pressure vessel nozzles and steam generator connections respectively. The housings of the main coolant pumps and of the primary isolating valves are also made of austenitic materials. The pros and cons of the different materials concepts are the subject of international expert discussions. Operating experience has shown that primary stresses in the main coolant lines can be kept low in both concepts. The degree of fatigue of the material is influenced by the mode of plant operation and by the routing of pipes. As a result of temperature stratifications and fluctuations, the connections of secondary pipes to the main coolant lines are frequently found to suffer from higher degrees of fatigue. This makes it advisable to put the welded joint at the materials interface in areas of low stress. To what extent this boundary condition is met in this case cannot be assessed, as no stress analysis of the primary system is as yet available in the necessary amount of detail.

According to the present state of the art, the findings derived from ultrasonic inspections of thick-walled austenitic welded joints and austenic-ferritic joints offer only a limited amount of information. Consequently, intermediate inspections are carried out in the fabrication process at various degrees of weld fill (radiography and surface crack tests). The results of these intermediate inspections are contained in the vendor's documentation, which has not been examined as yet. For this reason, no final statement can as yet be made about the quality of the thick-walled welded joints of austenitic components and the bimetallic welds which connect the pipe to the reactor pressure vessel and the steam generators.

In some areas, the surfaces of the welded joints are not plane enough for ultrasonic inspection techniques to be used to the necessary extent as in-service inspections. For this reason, the surfaces must be reworked. If any limitations to inspectability were to remain after that step, surface crack tests could be carried out from the inside, if necessary.

- Pressurizer

Operating experience from Units 1 to 4 has shown the material used for the pressurizer to be basically suitable. As its mechanical and engineering properties as well as reliability in fabrication are not yet known sufficiently well, the details of the data on the certificate and the manufacturer's documentation still need to be evaluated. If unsatisfactory toughness levels were found it would have to be assessed whether the pressurizer should be replaced or could be continued in use after some design modifications.

Design modifications are necessary for some types of nozzles. This is true, in particular, of welds with the roots not penetrating. The technical improvements necessary to remove test limitations are feasible. The higher loads possibly resulting at those points as a consequence of temperature stratifications and fluctuations can be limited by the appropriate mode of operation and by changes in pipe routing.

- Steam Generators

Operating experience from Units 1-4 has shown the material selected for the steam generator shell to be fundamentally suitable. As its mechanical and engineering properties as well as reliability in fabrication are not yet known sufficiently well, the details of the data on the certificate and the manufacturer's documentation still need to be evaluated. Any restrictions in testability can be removed satisfactorily by reworking the welds, modifying some nozzles and, if necessary, also some internals. Supplementary test findings can be derived from special techniques to be applied. Reference inspection of a steam generator of Unit 7 showed inadmissible findings. This gives rise to the requirement for supplementary ultrasonic inspection of the steam generators in Unit 5. At present, no conclusive statement is possible as to whether the transition from austenitic to ferritic components can be inspected sufficiently well. If necessary, some of the nozzle joints would have to be modified in design.

Operating experience with the steam generators of Units 1 to 4 has shown that water chemistry must be monitored by more stringent measures. Modifications to the condenser piping and changes in water chemistry management clearly could reduce the sensitivity of the steam generator tubes to local corrosive attack.

- Feedwater and Main Steam System

Operating experience in Units 1 to 4 has shown the unalloyed and low-alloyed, respectively, steel grades used in the tanks and pipes of the secondary system to be basically suited to this type of operation. As their mechanical and engineering properties as well as reliability in fabrication are not yet known sufficiently well, the details of the data on the certificate and the manufacturer's documentation still need to be evaluated. There is a danger of erosion-corrosion in some areas of adverse flow conditions. This limitation can be met largely by replacement of materials, local cladding, and by the appropriate type of water chemistry management (high-AVT mode). Surface crack inspections of welds of the tanks and pipes in category III have not been conducted so far and still need to be performed.

- Necessary Information and Verification

(1) A status report must be elaborated which represents the current state of knowledge about the reliability in fabrication and about the neutron exposure and corrosion characteristics of the 15Ch2MFA reactor pressure vessel steel.

(2) The mechanical and engineering data contained in the certificates cannot be allocated to any location of a specimen. In particular, the specimen shapes used for toughness tests have not been indicated. Supplementary information is required in this area.

(3) Some numerical information about mechanical and engineering data and chemical analyses respectively which differs greatly from the specifications must be checked. In addition, differences in the toughness data must be clarified which may have arisen from conversions of dimensions.

(4) More detailed information is required about the test procedure of austenitic-ferritic welded joints (especially about the forged-on nozzle necks of the reactor pressure vessel).

(5) The accommodation of loads from leakage accidents in the turbine hall must be verified in a supplementary procedure for the penetrations of the main steam and feedwater pipes through wall C.

(6) The verifications of the stability of primary systems components and their supports must be repeated on the basis of currently valid methods of calculation and, if necessary, supplemented by FEM calculations for specific load cases.

(7) To evaluate the concept of quality assurance used by the component manufacturers, the information available must be supplemented by inspections of the documented checks made during the construction, which are kept by the manufacturer. Non-destructive tests of the base metals must be verified and, if necessary, repeated.

(8) The result of the basic ultrasonic inspection of the base metal areas of the reactor pressure vessel (shell rings and bottom) must be presented, including the nozzles of NB 250.

(9) An inspection program (ultrasonic inspection from the inside, inspection by means of a TV camera from the inside and outside) for the nozzles and the area of holes in the top of the reactor pressure vessel must be presented.

(10) The consequences arising from the measured slant of the reactor pressure vessel with respect to the loads acting on the nozzle connections must be evaluated.

(11) For in-service inspections of the main coolant line and the connecting pipes of the pressurizer, a mechanized internal inspection (ultrasonic, visual inspection) must be provided. The test procedure used to inspect the longitudinal welds of pipe bends must be upgraded.

(12) An inspection method for austenitic-ferritic welds must be upgraded.

(13) The ultrasonic inspection of the steam generators must be repeated completely in line with the criteria in the KTA codes and standards.

(14) A concept of inspection of the steam generator tubes must be presented which also includes the areas of the bends.

(15) The defect in Unit 2 in 1982, which developed from a blind hole in the steam generator collector flange, must be examined for any conclusions resulting with respect to Unit 5.

(16) The validity of the radiographic inspections of the welds of main coolant pumps and main gate valves must be determined. If necessary, supplementary inspections must be carried out with optimized inspection techniques (test specimens).

(17) The vessel welds of the pressurizer must again be inspected ultrasonically for longitudinal and transverse flaws.

(18) Surface crack tests must be made for the welds of the vessels and pipes of the secondary system.

(19) The use of sufficiently leaktight condensers with pipes made of chrome nickel steel or titanium must be checked as a precondition for changing to the high-AVT mode of operation in the secondary system.

- Plant Upgrading

(20) The EOL fluence at the wall of the reactor pressure vessel must be limited.

(21) Special leakage monitoring systems must be provided for leak detection at the RPV top nozzles.

(22) Roots that are not penetration-welded, e.g. at spray nozzles and heater elements of the pressurizer and at NB 500 thermal buffers of the main gate valve, must be eliminated.

(23) Limitations to inspectability by non-destructive tests resulting from the test geometry or from excess weld metal must be eliminated.

(24) Accessibility for in-service inspections of the main coolant line and the connecting pipes of the pressurizer must be improved.

(25) An operational system must be installed for automatic monitoring of the water chemistry parameters in the primary and secondary systems.

(26) Resin catchers must be installed downstream of the special water treatment system (SWA 1 and SWA 1a).

(27) The pressurized area of the feedwater system must be protected against erosion-corrosion by suitable substitutions of materials or by claddings.

(28) Possibilities must be provided for inspecting the surfaces of welds on the secondary side of the steam generator collector.

5   Loads Resulting from Accidents

5.1   Analyses of Loss-of-Coolant Accidents and Transients

The accident analyses existing for design basis accidents in Unit 5 of the Greifswald Nuclear Power Station and other plants of the W-213 type were screened and evaluated in terms of completeness, sufficient conservativeness, comprehensibility, and plausibility of the results. The extent to which the design basis accidents are coped with by the existing engineered safeguards, by automatic actions of the safety systems and, if applicable (30 minutes after the onset of an accident), also by measures initiated manually, was examined. Where existing analyses or the inspectors' conservative estimates do not show beyond any doubt that the design basis accidents are coped with by those measures, additional analyses are required.

The existing analyses are evaluated on the basis of the technical safety codes and regulations derived from the German Atomic Energy Act. These are mainly the Safety Criteria for Nuclear Power Plants, the Accident Guidelines, the RSK Guidelines for Pressurized Water Reactors, and the statement made by the German Federal Ministry of the Interior about the single-failure criterion.

The evaluation was based on the Technical Project /1/ and on additional documentation (Addendum 19) /2/.

The safety report as part of the Technical Project /1/ contains data and findings not sufficiently well documented. Results of the design basis calculations can be reconstructed only in the rarest of cases. On the other hand, various project modifications were made only very recently and, for that reason, are not included in Addendum 19. Moreover, Addendum 19 deals with only a few selected accidents which, as in the Technical Project, are not documented satisfactorily. None of the reports contain a description of the computer codes used for the design basis calculations.

Consequently, no safety report is available which could meet today's requirements. Nor is this omission made up for completely by additional analyses performed by the architect engineer of the plant with computer codes of his own. Other accidents, which were analyzed within a special IAEA regional program for plants of the W-213 type in the Hungarian Paks Nuclear Power Station, can be applied to Unit 5 only to some extent.

The single-failure criterion was assumed in the existing analyses. Further restrictions of the availability of the safety systems acting for accident control were assumed in specific cases. The repair case required in German codes and standards was not taken into account systematically. In the majority of accident analyses, reactor scram is initiated by the first activation criterion. In the German codes and standards, failure of the first activation criterion must be assumed.

The findings and evaluations derived from examining the analyses are summarized below.

5.1.1 Loss-of-Coolant Accidents

The RSK Guidelines require for the design basis calculations of loss-of-coolant accidents that

- the calculated maximum fuel rod cladding tube temperature does not exceed 1200 °C,

- the calculated depth of oxidation of the cladding does not exceed 17% of the actual wall thickness of the cladding tube at any point,

- not more than 1% of all the zirconium contained in the cladding tubes reacts with steam,

- only small fractions of the core inventory (10% of the noble gases, 5% of the volatile solids, 0.1% of other solids) may be released into the containment. It has to be assumed that 10% of all fuel rods failed unless a lower percentage of failure can be demonstrated in a core damage analysis.

In addition, subcriticality and coolability of the reactor core must be ensured for long-term cooling after a loss-of-coolant accident.

- Leaks in the Primary System

At present, there is no comprehensive and sufficiently documented analysis of the double-ended break of the main recirculation pipe (2A-break) for Unit 5 of the Greifswald Nuclear Power Station. The existing results of computations do not demonstrate reliably that the above limits are observed. To demonstrate the adequacy of the design of the emergency cooling system, the whole sequence of the accident event, including the extent of damage, must be analyzed.

It must be pointed out that no computer code is at present available to analyze the extent of damage and determine the radioactive substances released from the fuel, which has been verified with data specific to fuel rod behavior in the WWER-440 reactor. Development work along these lines is necessary.

The break of an accumulator feed pipe leading into the annulus of the pressure vessel must be analyzed because, in this accident, the effectiveness of the emergency cooling system is restricted in particular by bypass flows to the leak.

Medium-sized and small leaks were the subjects of calculations in the Technical Project as well as in Addendum 19 to the Technical Project where a number of calculations of various leak cross sections (equivalent to NB 170, 113, etc.) was performed. The results elaborated with respect to leak cross section NB 113 are not plausible. It is expected that advanced thermohydraulic computer codes will determine accident sequences which will not give rise to core degradation. To assure this expectation, the case should be analyzed with an equivalent leak size of NB 113.

In the case of medium-sized and small leaks there is a pressure range, after the pressure accumulators have been discharged and until the low-pressure safety injection system is initiated, in which only high-pressure safety injection is effective. To support rapid depressurization in the primary system, fast automatic cooldown via the secondary system is thought to be necessary. It is expected that the break of the pipe connecting the pressurizer with the safety valves can also be coped with by the existing emergency core cooling systems in that case. An analysis backing this assumption must be performed.

Inadvertent opening of a pressurizer safety valve is coped with.

- Leaks from the Primary System to the Secondary System

The double-ended break of a steam generator heater tube is associated with a major release of activity to the outside unless short term manual measures are taken within the first 30 minutes after the onset of the accident (30-minute criterion). Without such measures, the defective steam generator can be filled up completely with water from the primary system. This will cause the opening of the atmospheric main steam dump (BRU-A) or of the steam generator safety valves. This entails the danger that the open valves, which are acted upon by liquid water, may fail in the open position.

The "double-ended break of a steam generator heater tube" accident requires detailed analyses to be performed, especially those from which automatic measures are derived to prevent inadmissible activity releases to the outside. One of the assumptions to be made in these analyses is the failure to close completely the main gate valves on the primary side. In addition, variants with and without the occurrence of the emergency power case must be studied. Some first thermohydraulic analyses of these open questions are at present being carried out by the architect engineer.

The break of the steam generator collector was not considered in the Project. After an accident in the Soviet Rowno Nuclear Power Station in 1982, analyses were carried out for the Paks Nuclear Power Station. No specific analyses of this accident exist for Unit 5. Without upgrading measures, this accident will be associated with an inadmissible release of activity (see Sec. 5.3). Detailed analyses are required to demonstrate the effectiveness of suitable upgrading measures.

5.1.2 Transients

- Reactivity Accidents

The analyses existing in the project documents of the inadvertent withdrawal of control elements were evaluated and found to be sufficiently conservative. These accidents do not result in an inadmissable load on the plant. The ejection of control elements requires supplementary analyses to be conducted with a three-dimensional reactor dynamics code.

A leak in the main steam system is not expected to cause recriticality of the reactor. However, to quantify the reactivity feedback, supplementary analyses must be carried out with 3D core models.

The existing analyses of reactivity effects in a double-ended break of the main coolant line and the inadvertent feeding of clean condensate to the primary system are sufficient and are deemed to be plausible. Recriticality is not expected to occur in these cases.

- Breaks in the Secondary System

According to the present design, the break of a feedwater pipe causes a reactor scram only after the water level in two steam generators has dropped by more than 400 mm (L_{DE} < -400 mm below the nominal level). Existing analyses show that unnecessarily large volumes of water are discharged through the leak before the reactor is shut down after some 80 s. In order to preserve the water inventory on the secondary side it is recommended to initiate reactor scram at an earlier point, when the level in only one steam generator has dropped to L_{DE} < -400 mm.

Supplementary analyses are required of the break and leakage respectively in a main steam pipe, in which the water entrainment phenomena on the secondary side are modeled as realistically as possible. The location and sizes of the leaks must be varied systematically to determine, in this way, the most adverse effects on the core inlet temperatures and the effectiveness of the different reactor protection criteria (release of interlocks HS-4 to HS-1). In case the basic safety of the pipes on the 14.7 m platform cannot be confirmed, analyses must be performed of the breaks of several main steam pipes. Some first thermohydraulic analyses of these open questions are at present being carried out by the architect engineer.

Analyses of the break of the main steam collector do not indicate whether, and how reliably, the "exceeding the pressure drop rate of 0.8 bar/s for at least 5 s" reactor scram criterion is reached. Further studies must be conducted until a suitable criterion can be defined.

- Failure of all Main Feedwater Pumps

As the analyses show reactor scram, even in a failure of all main feedwater pumps, occurs only when the "steam generator level low" criterion is reached in two steam generators or the "level very high" criterion is reached in the feedwater tank. To maintain the water inventory on the secondary side, it is therefore deemed to be necessary to immediately shut off both turbines and then initiate reactor scram by a directly responding criterion (e.g. "pressure low" in the high-pressure feedwater collector).

- Loss of Off-site Power and Operational Transients

Suffficiently extensive analyses exist of loss of off-site power and failure of the main coolant pumps, "defective heating in the pressurizer", "load rejection", and "failure of the main heat sink" as cases of operational transients. They can be interpreted to imply that, if the systems and protective devices function as designed, the plant will be transferred into a safe condition without any manual intervention.

- ATWS Accidents

No analyses are available of operating transients associated with failures of the reactor scram system (ATWS). In the RSK Guidelines, ATWS analyses are demanded for some selected operational transients.

5.1.3 Cold Water Strands

Cold water strands are important in assessments of the safety against brittle fracture of the reactor pressure vessel, when such cold water strands occur distributed asymmetrically across the annulus and at high pressure of the primary system. Cold water strands are of particular interest at that level of the core where the brittle fracture transition temperature from the base metal to the weld is enhanced by neutron exposure. No specific analyses of this subject exist for Unit 5 of the Greifswald Nuclear Power Station.

The existing analyses for Units 1-4 and those from international research projects respectively are not sufficient or cannot be extrapolated directly. It is recommended to perform analyses of the formation of cold water strands similar to the PTS Study /3/ conducted for the Loviisa Nuclear Power Station in Finland.

5.2 Pressure-resistant Compartment System with Pool-Type Pressure Suppression System, so-called Confinement System

The confinement system is a self-contained system of compartments surrounding the components of the primary system which are under high pressure and high temperature. It is made up of 44 pressure-resistant interconnected compartments. The confinement system also includes a pool-type pressure suppression system. To confine radioactive substances, a sufficient negative pressure is maintained in the confinement by means of the ventilation system in normal operation.

5.2.1 Basic Project Design Principles

When the confinement system of Unit 5 was designed, the safety criteria listed in the Technical Project /1/ were used as a basis. These criteria require installations for the retention and removal of radioactive substances (localization systems) to be provided which limit activity releases to permissible levels. The following detailed requirements apply:

- The localization systems must perform under accident conditions and must have sufficient capacity plus a reserve margin (i.e. redundancy).
- The entire primary system must be contained in the confinement.
- All penetrations through the outside walls of the confinement system must be fitted with double seals.
- Facilities must be provided for individual checks of the sealed penetrations at their rated pressure (corresponding to the design pressure of the confinement system of 145 KPa overpressure).

The design levels of maximum pressure and maximum temperature were determined in accordance with /1/ as the most adverse data obtained from accident calculations without any added safety margins. External impacts (earthquake, airplane crash, and explosion) were not taken into account in designing the confinement system.

5.2.2 Analysis of the Design Parameters of the Confinement System

To verify the design pressure and the design temperature, analyses were conducted under the assumption of a double-ended break of the main coolant line (NB 500) with a variety of boundary conditions. A multi-compartment computer model was used for the analyses.

According to the results of these calculations, the design values of the confinement system for pressure and temperature (245 kPa, 127 °C) are just about reached or slightly exceeded for this accident, depending on the boundary conditions.

The analyses show that the integrity of the deflection hoods in the pressure suppression pools plays a very important role with respect to the maximum accident pressure. Even the failure of only a small number of hoods per pool during the blowdown (>2 hoods) will cause the design pressure to be exceeded. The failure of 12 hoods per pool already gives rise to an accident pressure equivalent to the pressure which would prevail if there were no water in the entire pool-type pressure suppression system. The stability of the hoods must still be checked under various dynamic loads. However, insufficient strength of the hoods made of plastics has no decisive impact on the concept, as dampers made of an appropriate substitute material or a different geometry can be backfitted.

As only small or no safety margins exist and the failure of a few hoods will already lead to the design pressure and temperature levels being exceeded, supplementary calculations were carried out. They have indicated that increasing the size of the discharge area through the non-return valves into the air traps greatly reduces the peak pressure.

The design of the confinement was based on the assumption that 30 minutes after the onset of the double-ended break (2A) of the main coolant line a negative pressure relative to the environment would be restored in the confinement system as a result of steam condensation. It has not yet been possible to verify this design criterion, as no sufficient calculations are available about the long-term discharge behavior from the leak. More precise calculations are necessary in which the long-term heat input from the secondary system and various phenomena associated with condensation must be taken into account.

5.2.3 Differential Pressure Loads

To verify the design of the ceilings and walls of the confinement for possible differential pressure loads, no systematic analyses have as yet been performed of various break cross sections, break locations, and the resultant discharge rates. Also lacking are comparable data supplied by the manufacturer and operator respectively of the Greifswald Nuclear Power Station about the differential pressure levels used in the stress analysis of the building structure.

However, for a first check of stresses in the building structure it is possible to make use of the differential pressure data determined from the GRS analysis conducted to verify maximum pressure and temperature levels.

5.2.4 Dynamic Loads Acting on the Pool-Type Pressure Suppression System during Accidents

No verification was possible of jet forces acting on the baffle of the pool type pressure suppression system, possibly giving rise to loads which could not be accommodated. However, design modifications allowing jet forces to be absorbed would be possible without major restrictions.

No representative experiments exist to determine the maximum differential pressure expected to arise in accidents between the pools of the pressure suppression system and the hoods. Some first tentative calculations with DRASYS seem to indicate that the design level of 29.4 kPa will cover the differential pressures that are expected. However, detailed analyses still must be conducted to back this assumption.

During pool swelling the dynamic nature of the processes in the pools and the resultant fast pressure buildup can trigger the interlocking mechanism with the overflow dampers open. Complete closing of the dampers must be ensured in any case. If necessary, the design of the interlocking mechanism must be upgraded.

When steam-air mixtures are condensed in the water pool, pressure pulsations are generated. No design basis data exist about the resultant dynamic loads. As maximum vibrational loads caused by such condensation pulsations the pressure amplitudes are assumed which are derived from experiments, between +110 and -70 kPa at one vent pipe, and ±50 kPa at all vent pipes at the same time /4/. Accommodation of these pressure peaks by the steel structure of the pool-type pressure suppression system must be examined in a stress analysis with regard to material fatigue.

5.2.5 Jet Forces and Reaction Forces

It is not possible at present to assess the protection existing against jet forces and reaction forces originating in pipe breaks and also against any fragments ejected. It needs to be clarified which of the pipe systems within the confinement can be evaluated as being basically safe. In order to get a basic safety piping system, only a leak size of 0.1x the pipe cross section instead of a double-ended pipe break must be assumed as a basis on which to determine load magnitudes. To what extents fragments with a mass >5 kg need to be taken into account still requires a more detailed investigation of the possible formation of fragments. If required by subsequent detailed studies, possibilities seem to be available to ensure improved protection against consequential damage to adjacent systems by means of protective devices to be determined in each individual case. The limitations applying to accessibility, inspection and testing must be taken into account.

5.2.6 Leaktightness and Confinement Isolation

The leak rate at the design pressure of the confinement system amounts to approx. 0.6 vol.%/d, which is between the levels of the fully pressurized containment of a KWU pressurized water reactor (0.25 vol.%/d) and of the containment with pressure suppression system of the SWR-72 KWU boiling water reactor (1 vol.%/d). Consequently, based on the free volume, the leak rate corresponds to those of West German plants.

The confinement system includes a large number of pipe penetrations with shutoff valves (normally in triplicate or in duplicate with one check valve). The valves are powered by various emergency supply buses and are activated by redundant I&C

systems. The power supply trains and the activation trains are physically separate. On the whole, the basic principles of safe confinement isolation have been observed.

In the past, transfer locks and assembly openings frequently exhibited leakages, mostly for structural reasons. Weak spots in particular were the hatches above the steam generators. To ensure leaktightness of the building independent of any particular manual measures (such as resealing the doors of air locks), constructional modifications must be made.

After all leakages detected in the outside walls of the confinement system have been sealed off, the main contributions to the total leakage are likely to arise from the many pipe and cable penetrations. A leak extraction system at the penetrations could ensure controlled, filtered releases of leakages and partly compensate for the absence of an additional secondary containment structure.

5.2.7 Summary Assessment and Measures Required

A study was conducted to see to what extent the confinement system of Unit 5 meets the requirements in German safety-related guidelines in terms of design, construction, and function. In general, the data provided by the vendor and the operator of the plant about systems, geometric dimensions, plant operation, etc. were accepted without further verification.

Under normal operating conditions and accident conditions, the function of the confinement system corresponds to that of the containment. Verification indicated that the confinement system, as a containment with a pressure suppression system, in principle meets the requirements contained in the safety criteria and the RSK Guidelines.

However, the confinement system is not enclosed in a secondary containment structure, from which leakages could be collected for controlled discharge into the environment through filters. Such a safety enclosure, as required in the safety-related guidelines, therefore has not been achieved completely. The confinement system is not designed against external impacts (air plane crash, earthquake, and explosion).

Physical separation of the primary systems and engineered safeguards enclosed by the confinement system was not achieved in a consistent manner. To what extent these systems, and the confinement system, have been designed even against possible consequences of an accident, such as mechanical loads, still needs to be verified in detail.

The following questions of detail have not yet been studied in the evaluation carried out so far:

- Influence of the most adverse operating status in the primary system upon the maximum pressure and maximum differential pressures in the confinement system.

- Dynamic loads arising from condensing steam in the water of the pool-type pressure suppression system.

- Jet forces and reaction forces, behavior of fragments.

- Measures against consequential damage arising to electrical systems and I&C systems.

- Hydrogen generation after loss-of-coolant accidents.

- Measures Required

A number of recommendations and measures were derived from the investigations. They relate to proposals for further studies and the necessary backfitting measures.

(1) Verification of the integrity of the plastic deflection hoods in the pool-type pressure suppression system under accident conditions, with aging taken into account.

(2) Verification of the accommodation capability of dynamic loads by hoods, pools and building structures during condensation processes in the pools of the pool-type pressure suppression system.

(3) Detailed studies of pressure buildup and differential pressures in the confinement system.

(4) Detailed study of the effectiveness of the sprinkler system in the light of failure criteria.

(5) Doubling the discharge area through the non-return valves into the air traps.

(6) Prevention of consequential damage caused by jet forces, reaction forces, missiles, thermal loads, and moisture.

(7) Leakage extraction system at all penetrations and leakage points detected.

(8) Improved sealing of leakages at the air locks.

(9) Reliable activation of the reactor scram prior to the actuation of the rupture disc between the steam generator box and the shaft of the pool-type pressure suppression system responds.

5.3 Radiological Impacts

In the Accident Guidelines, calculations of potential radiological consequences are required for a number of radiologically relevant design basis accidents to verify observance of the accident planning levels listed in Sec. 28, Subsec. 3 of the Radiation Protection Ordinance. Methods of calculation must be employed which are defined in special accident calculation principles.

Such calculations were carried out for the following accidents:

- Double-ended break of a main coolant line.

- Fuel assembly damage during handling.

- Break of the steam generator collector.

An accident in western-type pressurized water reactors (break of the steam generator tube sheet) which would correspond to the break of the steam generator collector is not includeed in the Accident Guidelines as a design basis accident. The potential radiation exposures determined as radiological consequences were compared with the accident planning levels.

At a distance of about 1.5 km around the Greifswald Nuclear Power Station, a protected area has been defined with restrictions of sojourn and use. These restrictions have not been taken credit of in the calculations performed here.

5.3.1 Loss-of-Coolant Accidents

In accordance with the requirements in the RSK Guidelines, it must be demonstrated that not more than 10% of the fuel rod cladding tubes become defective in a loss-of-coolant accident. For this verification, studies must be conducted of the rupture behavior of the cladding tubes and extensive emergency cooling analyses performed to determine the cladding tube temperatures arising in the course of an accident. For the Greifswald Nuclear Power Station and the fuel assemblies used in it, no sufficiently documented calculations exists from which the above criterion could be seen to be met.

To determine the radiological consequences of a loss-of-coolant accident it is therefore assumed, pessimistically, that all fuel rod claddings will turn defective. However, it is important to point out that the fraction of fission products released from the core will increase not proportionally with the number of the ruptured fuel rods, but clearly less; see /5/.

The activity release into the environment associated with the accident is determined by the duration for which an overpressure exists in the confinement system. The leak rate of the confinement system at design pressure is approx. 0.6 vol.%/d (see Sec. 5.2.6). The duration of the overpressure phase is determined greatly by the effect of the spray system. As a consequence, a negative pressure will be reestablished in the confinement system in the longer term in the course of the condensation processes.

The design of the confinement system was based on the assumption that, 30 minutes after the onset of a double-ended (2A) break of the main coolant line, a negative pressure relative to the environment would be restored in the confinement system. Although it has not been possible so far to verify this design criterion, it has been used as a basis in estimating the activity release associated with the accident.

In analogy with the basic data underlying the calculation of the accident, it was assumed that of the airborne iodine released 10% would be elemental iodine and 90% would be iodine aerosols.

For the potential radiation exposures, the contributions coming from ingestion, inhalation, and external irradiation (clouds, ground) were determined. For an infant as the critical person and the thyroid as the critical organ, all maximum exposures are encountered at a distance of 320 m (from the stack). The resultant thyroid dose is 56 mSv (limit, 150 mSv), and the effective dose is 9.5 mSv (limit, 50 mSv). For adults, 20 mSv is found for the thyroid, and 7.7 mSv for the effective dose, with the same limits.

The potential radiation exposures determined thus do not exceed one third of the accident planning levels.

5.3.2 Fuel Assembly Damage during Handling

For the handling accident, a fuel assembly crashing and being damaged during refuelling, it is assumed that the outer fuel rods in two adjacent corners will become defective in the hexagonal fuel assembly (see Sec. 4.1.4). The radioactive noble gases emanating from the defective fuel rods will be released into the environment by the ventilation system right through the stack. The iodine discharged by defective fuel rods enters the pool water. However, a small part of it may change into the gas phase in the long run and be released into the atmosphere.

The radiation exposures associated with various exposure pathways were determined. The maximum levels for an infant as the critical person and the thyroid as the critical organ are encountered at a distance of 2000 m (from the stack) for ingestion, at 500 m for inhalation and external irradiation from the cloud, and at 360 m for external irradiation from the ground. The total thyroid dose is 27 mSv, as compared to the limit of 150 mSv, and the effective dose is 0.86 mSv, as compared to the limit of 50 mSv. For adults, the thyroid dose was found to be 7.4 mSv and the effective dose, 0.24 mSv, always with the same limits. On the whole, the calculated potential radiation exposures for the accident considered are clearly below the accident planning levels.

5.3.3 Break of the Steam Generator Collector

The break of the steam generator collector is associated with an activity release into the environment.

In order to limit the release, manual measures must be taken at short notice to turn off the defective steam generator. No analysis specific to Unit 5 exists of the thermohydraulic sequence of events and the radiological consequences of this accident. For this reason, a simple estimate of the radiological consequences is made for a selected case.

A leak of 80 cm² is assumed to have been caused in the steam generator collector by a break of the collector top. No manual measures are taken into account for the first 30 minutes.

For an infant as the critical person and the thyroid as the critical organ, the peak radiation exposures for inhalation, ingestion and external irradiation from the cloud and from the ground are all at 280 m (distance from the stack). The thyroid dose of 230 mSv exceeds the limit of 150 mSv. The effective dose of 15 mSv is below the limit of 50 mSv. For adults, thyroid doses of 76 mSv and effective doses of 10 mSv are found to give rise to no levels exceeding the same limits.

The radiation exposure of the critical organ (thyroid) of the critical group of persons (infants) calculated under these conditions for the break of the steam generator collector top as the representative accident exceeds the corresponding accident dose limit of 150 mSv by roughly a factor of 1.5.

Further studies must be conducted of this accident. If necessary, backfitting measures must be taken in order to reduce the activity release associated with this accident.

References, Section 5

/1/ Teploelektroprojekt, Technisches Projekt (Technical Project), Nord III/IV Nuclear Power Station, Moscow, 1974

/2/ Zusätzliche Technische Dokumentation, "Nachtrag 19" (Additional Technical Documentation, "Addendum 19"), Moscow, 1985

/3/ H. Tuomisto, Thermal-Hydraulics at the LOVIISA Reactor, Pressure Vessel Overcooling Transients, Helsinki, 1987.

/4/ KKP 1, Gutachterliche Stellungnahme zu den Belastungen des Sicherheitsbehälters mit Druckabbausystem (Comment by the Expert Consultant about the Loads Arising in the Containment Equipped with a Pressure Suppression System), Technischer Bericht Nr. 116-528-6.3.20, TÜV Baden e.V., 1978.

/5/ F. Lange, H. Friedrichs, W. Ullrich, J.P. Hosemann, Neuere Analysen des Spaltproduktverhaltens nach einem Kühlmittelverlust-Störfall (Recent Analyses of Fission Product Behavior Following a Loss-of-Coolant Accident), atomwirtschaft/atomtechnik, XXVII, No.2, 1982.

6 Systems Engineering

6.1 Systems Analysis of Process Engineering Aspects

Systems engineering aspects are evaluated on the basis of an evaluation of accident sequences. In the selection of the initiating events the Accident Guidelines, specific plant conditions, and experience accumulated in the commissioning phase of Unit 5 so far are taken into account.

The yardstick by which safety-related design is evaluated are the regulations under the Atomic Energy Act valid in the Federal Republic of Germany, especially the Safety Criteria for Nuclear Power Plants, the RSK Guidelines for Pressurized Water Reactors, the Accident Guidelines, and the KTA Codes.

This also means that single failures and repair cases are included in determining the required redundancies of engineered safeguards systems.

Where valid codes and regulations are not met, investigations are performed to see

- whether such deviation gives rise to a deficit in safety;
- which measures can be taken to make up for a safety deficit by other existing devices or properties of the plant.

6.1.1 Initiating Events

The study is restricted to events which may result in damage to the reactor core. Among the initiating events, it distinguishes between events leading to loss-of-coolant accidents and events leading to transients. For each initiating event, the simultaneous occurrence of the loss of off-site power has also been investigated. After a loss of off-site power, additional criteria for reactor scram exist (failure of the last turbine, failure of more than three main coolant pumps). Moreover, after the loss of off-site power, the control rods will be dropped into the core when the voltage at the feed ends of the reactor scram system fails for more than 1.5 s.

- Loss-of-Coolant Accidents

In addition to large, medium-sized and small leaks of the primary circuit, leaks of the pressurizer, leaks of one or more steam generator heater tubes, and leaks of pipes connected to the primary system which extend outside of the confinement system are also analyzed.

- Transients

The following transients are examined:

- Loss of off-site power.
- Loss of main heat sink.
- Loss of feedwater supply system.
- Overfeeding of the pressurizer.
- Load rejection by the turbines.
- Leak of a main steam pipe.
- Leak of the main steam collector.
- Leak of a feedwater pipe.
- Leak of a feedwater collector.
- Failure of the service water system (NKW-A) and the component cooling system (NKW-B).
- Reactivity transients.
- Startup and shutdown events.
- Anticipated operational transients with failure to scram (ATWS).

Only those sequences of events are treated below, the evaluation of which indicated defects, and from which proposed improvements were derived, respectively. The defects of a system are addressed only once, i.e., they are not repeated in the further course of the discussion of other initiating events.

Loss of off-site power, the reactivity transients, and overfeeding of the pressurizer do not give rise to any further conclusions with respect to systems engineering.

6.1.2 Event Sequences in Loss-of-Coolant Accidents

6.1.2.1 Large Leak (NB 200 to NB 500)

Reactor scram and the startup of the emergency cooling chain (emergency cooling system and its auxiliary systems) are triggered off by diverse criteria (pressure drop in the primary system, decrease of water level in the pressurizer, pressure increase in the confinement system).

The emergency cooling criteria cause the Diesel generators to start up and the emergency cooling chain to be switched sequentially to the Diesel emergency power supply systems.

The reactor shuts itself off automatically through the negative void coefficient of reactivity. The core is held subcritical by feeding boric acid solution from the accumulators and then from the emergency cooling storage tanks. The four accumulators feed directly into the reactor pressure vessel as soon as the pressure in the primary systems drops below 5.4 MPa.

Emergency cooling is continued by high-pressure and low-pressure injection. After the boric acid storage tanks (65 m³ each of 40 g/kg) have been emptied, the high-pressure emergency cooling pumps automatically switch to the suction side of the low-pressure emergency cooling pumps. After the storage tanks of 500 m³ each have been emptied by 75%, the high- and low-pressure emergency cooling pumps and the sprinkler pumps are automatically switched to sump recirculation. In the sump recirculation mode, the residual heat is removed by the emergency core coolers and the residual heat removal system cooled directly with sea water.

- Defects Recognized and Upgrading Measures Required

(1) The reliability of feeding from the accumulators into the reactor must be improved. The existing switching system must be modified. The possibility must be analyzed of the accumulator shutoff ball failing, or one of the two isolating valves wrongly closing.

(2) To prevent an overpressurization of the accumulators, the tightness of both check valves in the accumulator connecting pipes must be monitored.

(3) The reliability of the position indication of the shutoff balls in the accumulators must be improved. Otherwise there is danger of not recognizing an inadvertant closure of the accumulators.

(4) When actuated, power is supplied to the emergency cooling chain only by the emergency power Diesel system. The emergency cooling systems should be supplied from the emergency power Diesel generators only if the station service power supply has failed (see also electricity supply, Sec.6.2).

(5) When required, a motor-operated valve must open in each pump discharge line. These motor-operated valves must be replaced by check valves with monitoring of the pipe section between the valves. Additional isolating valves must be provided for repair cases.

(6) The low-pressure emergency coolers are cooled directly with sea water (missing activity barrier, contamination hazard of the low-pressure emergency coolers for long-term heat removal). Backfitting a three-leg component cooling system is necessary.

(7) Two series-connected isolating valves must be opened to operate the residual heat removal system. To ensure reliable opening, another valve group connected parallel to the existing isolating valves would be useful. The group of valves must be monitored for internal leakage.

(8) For emergencies, using the sprinkler pumps to replace failed low-pressure emergency cooling pumps for residual heat removal is considered to be useful. The reliability of possible technical solutions must be investigated.

(9) The redundancy and reliability of the water level sensors in the three pump compartments of the emergency core cooling system and residual heat removal system must be improved.

(10) The partitions separating the pump compartments must be checked.

(11) If the pipe connecting the hot and the cold legs of the main coolant line is necessary to prevent water blockages during leakage accidents (which remains to be examined), the valves in the connecting pipe should be permanently in the open position. (In Units 7 and 8 no valves have been planned for the connecting pipe.)

(12) The monitoring capability of the isolation of the building drainage system connecting the pump compartments, which does not exist as of now, must be backfitted.

6.1.2.2 Medium-Sized Leak (NB 25 to NB 200)

Reactor scram is initiated by the same criteria as in a large leak. High-pressure injection is required to cope with the accident.

If, at a primary system pressure equal to or higher than 12.2 MPa, the pumps are operated in the minimum discharge mode, the water in the 65 m³ boric acid storage tank is heated at approx. 5 K/h by the pump waste heat.

Adding a flow restrictor allows the high-pressure emergency cooling pumps to be operated into the low-pressure range.

- Defects Recognized and Upgrading Measures Required

(1) For cooling the bearings of all three high-pressure emergency cooling pumps, the single-leg component cooling system (NKW-B) is used. In principle, it is possible to switch manually to the component cooling system of the main coolant pumps, but this will hardly be feasible under accident conditions.

(2) If it becomes necessary to heat the boric acid solution in the storage tank in order to guarantee resistance to brittle fracture, cooling down the minimum water discharge of the high-pressure emergency cooling pumps must be ensured by the component cooling system to be newly installed.

(3) In case a sump return line is plugged up, the water inflow to the other two sumps must be ensured (connection among the three building sumps).

(4) The high-pressure injection lines of the emergency core cooling system and the residual heat removal system and the injection line of the volume control system have no pipe whip restraints. It should be investigated whether consequential failures could arise from pipe leaks.

6.1.2.3 Small Leak (<NB 25)

Reactor scram (HS-1) follows the same criteria as in a large leak. However, it is doubtful whether a scram is activated by the pressure rise in the confinement system.

As long as off-site power is available, the plant is cooled through the operational systems on the secondary side. Additional depressurization in the primary system is achieved by the pressurizer spray system. If off-site power is lost, the emergency feedwater system and the atmospheric main steam dump station (BRU-A) are required for residual heat removal. It is not possible to use the pressurizer spray system if only emergency electric power is available. Pipes of <NB 80 in the confinement system have not been routed systematically in the project. This fact adds to the number and length of pipes of small diameter (additional vent and drain pipes) and, hence, to the expected frequency of small leaks.

- Defects Recognized and Upgrading Measures Required

(1) In case of an accident, some confinement isolation valves should have a capability for reopening after the building has been isolated and the leakage detected. This applies, e.g., to valves in the injection line of the volume control system (feeding system), which would offer an additional feeding possibility in an emergency core cooling case.

(2) An emergency power supply of the feed pumps of the volume control system should be provided (to be used, e.g., for pressurizer spraying).

(3) Pipes <NB 80 in the confinement system must be run in accordance with the project design.

6.1.2.4 Pressurizer Leak

A distinction must be made between a leak entailing a loss of coolant into the confinement system and a leak produced by an erroneously open pressurizer safety valve. As a consequence, steam enters the pressurizer relief tank. After the rupture membrane of the relief tank has responded, the primary coolant will flow into the shaft of the pool-type pressure suppression system.

If the pressurizer safety valve remains erroneously open, only one reactor scram criterion (low pressure in the primary system) is available. The coolant discharged from the pressurizer relief tank collects in the sump of the pool-type pressure suppression system. If the rupture membranes fail to open the water cannot flow over. In that case, only one leg of the emergency core cooling system and residual heat removal system is available for long-term emergency cooling.

- Defects Recognized and Upgrading Measures Required

(1) The pressurizer safety valves must be demonstrated to function for the passage of steam-water mixtures and water.

(2) A pressurizer relief valve with preclosing capability must be backfitted. The actuation pressure must be set at a level below the response pressure level for the fluid-actuated steam control valves of the safety valves.

(3) The pressure buildup in the pool-type pressure suppression system and in the confinement must be determined for the "pressurizer safety valve stuck open" accident. If necessary, pressure sensors must be backfitted in the shaft of the pool-type pressure suppression system.

(4) For the "pressurizer safety valves stuck open" accident the diverse actuation for the "opening of a pressurizer safety valve" reactor scram case must be backfitted.

(5) The local arrangement of the pressure sensors for reliable pressure measurement in the confinement system needs to be checked.

6.1.2.5 Leakage of a Steam Generator Tube

Reactor scram initiated by the "pressure drop in the primary system" and "filling level decrease in the pressurizer" criteria is delayed or prevented because of the low leak rate. A number of readings allow the plant personnel to detect the accident, scram the reactor, identify the faulty steam generator and isolate it first on the primary and then on the secondary sides. The accident can be controlled with the existing operational systems.

If the plant personnel fails to initiate a reactor scram, there will be load rejection of the turbine, reactor scram, followed by turbine tripping plus the actuation of the emergency cooling chain and of the atmospheric main steam dump station (BRU-A). In case one of the two main isolating valves (HAS) should fail to close completely, the pressure in the primary system would have to be reduced quickly below the actuating pressure of BRU-A. There are various possibilities of doing this.

In case of loss of off-site power, the condenser steam dump station (BRU-K) is not available for heat removal. In this case, therefore, manual actions are required early in order to prevent activity from being discharged into the atmosphere through BRU-A.

The primary system should be cooled down through the BRU-A of a steam generator that is not affected.

In order to prevent prolonged opening of the BRU-A of the faulty steam generator, the blowdown valves to the flash tank are kept open until the primary isolating valves have been aligned manually.

- Defects Recognized and Upgrading Measures Required

(1) The "increase in main steam activity" criterion should be included as an additional criterion activating reactor scram.

(2) Accidents with leaks (e.g. leak in the steam generator collector) from the primary system into the secondary system must be demonstrated to remain within the accident planning levels. Possibilities should be examined, and, if necessary, ensured to improve tight closing of the main coolant lines by the

main gate valves, without subsequent manual retightening, as well as reliable reclosing of the main gate valves under the full differential pressure.

(3) Electric power is supplied to the main gate valves by the emergency power Diesel units in case the emergency cooling system is challenged or in the case of loss of off-site power. When establishing the emergency power balance, the power consumption of the motors driving the main gate valves must be taken into account.

6.1.2.6 Leakage of Several Steam Generator Tubes or Leakage of the Steam Generator Collector

In a rupture of the collector top, a leak of about 80 cm² arises between the primary and the secondary system, which corresponds to a double-ended break of 29 tubes. Such leaks will cause the reactor scram system and the emergency cooling chain to respond immediately. The manual measures to be taken in principle correspond to those taken in managing a rupture of a single steam generator tube leak. Measures must be provided in order to meet the accident planning levels (no unadmissable radioactive release into the environment).

- Defects Recognized and Upgrading Measures Required

(1) It needs to be investigated to what extent leak cross-sections larger than 80 cm² in the steam generator collector can be excluded and whether the leak-before-break criterion applies.

(2) Measures must be taken on the primary and secondary sides to establish automatic temperature and pressure reductions for the primary system, which will work also in case of loss of off-site power.

(3) In case of a leakage in a steam generator tube or a collector and simultaneous failure of a main gate valve, overfeeding of the defective steam generator by the high-pressure emergency cooling pumps must be prevented by automatic measures.

6.1.2.7 Leakage in a Pipe Connecting to the Primary System outside the Confinement System

In such a leak, the coolant discharged will not enter the sump and, consequently, can no longer be used for emergency cooling. All pipes penetrating the confinement system are equipped with several confinement isolation valves. Low-pressure systems are separated from the primary system by double isolations.

Leaks through defective heat exchangers into the component cooling systems can be recognized by increases in level and by the activity in the compensation tanks.

- Defects Recognized and Upgrading Measures Required

(1) It must be investigated whether pipes, confinement isolation valves, and pipes between the confinement system and the confinement isolation valves have been designed for the primary system pressure. Possible pressure waves must be taken into account.

(2) It must be investigated whether consequential damage could arise to confinement isolation valves and pipes.

(3) The safety valves on the component cooling water side of the heat exchangers must be designed at least to withstand the break of a heat exchanger tube. In case of an activity increase on the component cooling water side, the confinement isolation valves and the isolating valves on the primary side of the heat exchangers must be closed automatically.

6.1.3 Sequences of Transient Events

6.1.3.1 Loss of Main Heat Sink

Loss of main heat sink may be caused by failure of the main cooling water, loss of condenser vacuum, failure of the main condensate system, turbine failure, and failure of the condenser steam dump (BRU-K) to open. Loss of main heat sink will be described briefly by the example of a failure of the main cooling water:

Failure of the main cooling water initiates vacuum protection of the turbine condensers, tripping of both turbosets, and reactor scram. The atmospheric main steam dump control valves (BRU-A) are actuated, releasing live steam into the atmosphere until the pressure transient has been dissipated.

After the main steam temperature and the main steam pressure have been reduced, the residual heat is removed through the cooldown system which, however, is not supplied with emergency power and whose safety valves must not come into contact with water. If the cooldown system is not available, the residual heat is released into the atmosphere through BRU-A over extended periods of time.

- Defects Recognized and Upgrading Measures Required

(1)   The cooldown system must be supplied with emergency power.

(2)   The safety valves of the cooldown system must be designed for water blowdown.

6.1.3.2 Loss of Main Feedwater

Loss of the main feedwater leads to a number of signals in the main control room. The operator can manually add the startup and shutdown pumps feeding warm water from the feedwater tank. If the steam generator filling level continues to drop, reactor scram and emergency feeding of cold water are initiated. If the operator is late in taking manual measures, reactor scram is initiated with a delay, and the pressurizer safety valves may be actuated. The pump for startup and shutdown is not supplied with emergency power, is not interlocked, and has no redundancy. The three redundant pumps of the emergency feedwater system are supplied through one common intake pipe from a 1000 m³ tank.

- Defects Recognized and Upgrading Measures Required

(1)   A "pressure rise in the primary system" criterion for actuating reactor scram must be introduced.

(2)   No automatic start of the startup and shutdown pump when the filling level in the steam generators drops and no connection of the pumps to the

emergency power system. This deficiency must be eliminated in connection with a new feedwater concept (see Sec. 6.1.3.6).

(3) Introduction of a turbine power limitation and reactor scram respectively as a function of the number of failed main feedwater pumps.

(4) Installation of position indicators in the main control room for the bypass valve of the high-pressure preheater.

6.1.3.3 Failure of Turbo Generators

In normal operation of the control systems, no reactor scram will be actuated if one turboset fails. Failure of both turbosets causes reactor scram.

However, this interlock can be short-circuited by a switch in a room next to the main control room.

The control system is designed for full-load rejection to the station service load level.

- Defects Recognized and Upgrading Measures Required

(1) The "failure of the last operating turboset" criterion for actuating reactor scram can be rendered inoperative by an easily accessible switch. This interlock must be fully automated.

6.1.3.4 Leak of a Main Steam Line

A main steam line leak in the confinement system causes the pressure in the confinement system to rise, reactor scram to be initiated, and all steam generator purging valves to be closed. The quick-acting isolating valve closes the main steam line affected. In the same loop, the main coolant pump (MCP) is switched off and the feedwater supply to the steam generator (SG) is stopped. If the quick-acting isolating valve fails to close, the check valve in the main steam line prevents the other steam generators from feeding the leak through the collector. As the steam generator is boiling out, the coolant temperature in the primary system drops considerably. However, no recriticality will arise. If the pressure in the confinement system were to rise to a level above 0.01 MPa, the emergency cooling chain would be actuated.

If there is a leak in the main steam line right upstream of the turbine, it is doubtful whether the criteria initiating reactor scram and leakage isolation will be attained. However, the turbine affected will be shut off by the turbine protection system.

- Defects Recognized and Upgrading Measures Required

(1) The activation of the steam generator safety valves must be redundant, and the safe opening and subsequent reliable closing functions must be verified including a reliability analysis.

(2) The installation of a steam generator safety valve with a lower actuating pressure, which can be triggered and which is equipped with an isolating valve, and the installation of an isolating valve upstream of BRU-A is required; however, 100% steam dumping capacity must be ensured through safety valves which cannot be isolated.

6.1.3.5 Leak of the Main Steam Collector

If there is a leak in the main steam collector, it is doubtful whether the "depressurization rate in main steam collector" criterion for actuation of reactor scramming will be reached.

- Defects Recognized and Upgrading Measures Required

(1) The response level for the "depressurization rate in main steam collector" reactor scram criterion must be examined.

6.1.3.6 Leak of a Feedwater Pipe

If there is a leak between the steam generator and the check valve, the escaping steam causes the pressure in the confinement system to rise. If the leak is located in the inaccessible part of the confinement system, the pressure rise causes a reactor scram. If the leak is in the accessible part of the confinement system, no reactor scram will be initiated, as no pressure sensors are installed in that area. If the filling level in the steam generator drops by 110 mm, the valves on the discharge side of the startup and shutdown pump open (feeding through the emergency feedwater system, low-power control). The operator must start the pump; if the leakage cannot be compensated, a drop in the filling level of at least two out of six steam generators will cause a reactor scram.

A leak in the turbine hall causes the filling level in the feedwater tank to drop and, subsequently, the feedwater pumps associated with the turbine to fail. If the operator does not initiate reactor scram, the reactor will be scrammed by the drop in the filling level of the steam generators. Consequential damage can arise both to the operational systems and to the electrical systems supplying the emergency feedwater pump drives.

- Defects Recognized and Upgrading Measures Required

(1) An independent emergency standby feedwater system must be installed. This system must be protected against impacts spreading in the plant (flooding, fire, turbine explosion) and external impacts alike.

(2) The present emergency feedwater system must be connected to the feedwater tanks.

(3) Additional possibilities of feeding emergency feedwater must be created (e.g. connecting nozzles for accident management).

(4) The ferritic purging lines and emergency feedwater lines have no pipe whip restraints even within the confinement system. It must be examined whether whipping restraints are required.

(5) A position indication of the valve in the high-pressure preheater bypass line must be installed in the main control room so that there is a possibility to check the feedwater supply in case of failure of the high-pressure preheater column.

6.1.3.7 Leak of a Feedwater Collector

If there is a leak in the feedwater intake collector, the content of the feedwater tank will flow into the turbine hall, as it cannot be stopped by motor-driven valves. The collector is split up into two half sections by valves, with the consequence that only two pumps will be affected directly. The pumps are shut down by protective interlocks.

Turbine trip is achieved by means of the "pressure decrease in the main steam system" signals. After shutdown of the last turbine, or a drop of the filling levels in two steam generators, the reactor is scrammed. Consequential damage in the turbine hall may arise to the electric drives of the emergency feedwater pumps, among other systems.

If there is a break in the upper feedwater collector, consequential damage can arise to the main steam pipes and to a emergency feedwater line on the 14.7 m platform.

- Defects Recognized and Upgrading Measures Required

(1)   Technical solutions must be elaborated to prevent consequential damage, arising from a leak in the area of the 14.7 m platform, to other pipes or pieces of equipment in adjacent areas of the compartment.

(2)   Motor-driven isolating valves with position indicators in the main control room must be installed in the feedwater intake pipes instead of the planned isolation by blanks.

6.1.3.8 Failure of Main Cooling Water System and Service Water Systems

The main cooling water system and the service water systems can fail as a result of a blockage of the water inlet or of flooding of the intake structure, e.g. as a consequence of a break in the discharge canal. Failure of the main cooling water systems leads to a loss of the condenser vacuum, thus causing turbine trip and reactor scram.

The residual heat must be removed by way of the atmospheric main steam dump station (BRU-A), as neither the cooldown system nor the emergency cooler are available. Both are cooled by the service water system (NKW-A). In order to cool the reactor down, accident management measures must be taken. There is enough time to initiate such measures.

- Defects Recognized and Upgrading Measures Required

(1)   Flooding of the intake structure of the service water system NKW-A so that all redundant systems are involved must be prevented by appropriate measures.

6.1.3.9 Startup and Shutdown Events

Many manual measures must be carried out during startup and shutdown, among them safety-related interlocks which must be activated and deactivated. The plant personnel have check lists for these steps.

The switching status and the switching actions completed are checked by means of the instrumentation in the control room (mimic diagrams, displays, circuit diagram pinboards, records of switchings). No automatic procedures exist for startup and shutdown, nor is it possible to verify automatically the switching status and the positions of interlocks.

- Defects Recognized and Upgrading Measures Required

(1) The switching status and the set point vs. actual value comparison of the interlock positions must be monitored automatically.

(2) An automatic system must be installed which ensures shutdown reactivity in any operating status.

(3) In the neutron flux measuring system, adjustments of the measuring chambers, the switches changing measuring ranges, and power matching of the "neutron flux $\geq$110% of permissible reactor power" scram criterion must be automated.

(4) When revising the operating manuals, the startup and shutdown procedures must be described more precisely.

6.1.3.10 ATWS Accidents

Unit 5 has not been designed to cope with ATWS accidents. There are no accident analyses in which failure of the reactor scram system is assumed.

There is no secondary reactor shutdown system. The operational make-up system does not meet the criteria of a secondary reactor shutdown system (emergency power supply, pump head, feed rate).

- Defects Recognized and Upgrading Measures Required

(1) A redundant secondary reactor shutdown system must be installed.

6.1.4 Summary

The design in terms of reactor physics and thermal engineering of the advanced WWER-440/W-213 line largely corresponds to the design of the older WWER-440/W-230 line. In the technical assessment of the systems engineering side of the plant, the favorable properties of the WWER-440 were taken into account, such as the relatively low power density, the attenuated xenon oscillation characteristics, the isolation capability of the main coolant lines, and the large water volumes in the primary and secondary systems.

Shortcomings specific to the WWER-440 line, especially in the areas of electrical and instrumentation systems and control engineering as well as process engineering in the turbine hall, such as the physical concentration of all main steam and feedwater pipes on the 14.7 m platform, also exist in Unit 5. Although the design in three legs, largely without meshing and physically separate, of the safety systems has been greatly improved, there are still conceptual weaknesses and deficiencies in the construction of the components, which need to be eliminated.

- These are the most important upgrading measures:

- Installation of an autonomous emergency standby system consisting of a steam generator emergency feed system, an additional boration system for diverse reactor shutdown, a reactor protection system, and an emergency control room.

- Installation of a redundant nuclear component cooling system serving, among other purposes, to cool the emergency coolers as well as the HP emergency cooling pumps.

- Connection of the three sumps of the confinement system.

- Upgrading the pressure protection devices for the primary and secondary systems with controlled depressurization taken into account.

- Eliminating oil leakages in the operation of the main coolant pumps.

- Protecting pipes and equipment in the main steam and feedwater systems on the 14.7 m platform against spreading impacts (protection against consequential failures, fire, turbine explosion).

- Rerouting the small pipes in the confinement system.

The shortcomings in design and, to some extent, the upgrading measures required are listed in greater detail after the discussions of the respective accident situations. Planning and preparing for each individual measure requires detailed investigations to be carried out to see whether these measures might impair the engineered safety features of the entire plant.

6.2 Electric Power Supply

- Grid Connection

The unit, with its two unit generators, has been designed for interception and maintenance at the station service power in case of a grid failure. The circuit design of the grid connections and the station service power supply correspond to the basic requirements. As the two main connections to the grid and the standby connection are supplied from a 220 kV outdoor switchyard, there is still the possibility of failure (e.g. as a result of the destruction of the outdoor switchyard) of all three grid connections. For this reason, an improvement of the supply situation on the grid side is deemed to be necessary.

- Station Service System

No basic objections are raised to the circuit concept of the station service and standby switching systems, including the existing interconnections between the two units.

As a consequence of defects in the automatic switchover system for standby grid supply, a revision is thought to be necessary in the light of existing voltage and current conditions. If the main coolant pumps in a special switching status are supplied through the 6 kV standby distribution systems, failure of four pumps as a result of one short circuit in a bus is possible.

- Emergency Power System

The emergency power system meets all conceptual requirements with respect to its circuit design and physical and functional separation. The power balances shown indicate that the Diesel systems have practically no power reserves. In addition, the method of balancing the power for the emergency power system is not in line with the requirements listed in the KTA Code. Consequently, power balances according to KTA must be established for all accidents to be taken into account.

The automatic programs for starting and connecting the emergency power Diesel systems and also for the sequenced connection of emergency power loads (SAOS/GZ) do not meet requirements. The following are the main shortcomings:

- Activation in case of underfrequency is missing.
- Activation level in case of undervoltage is set too low.
- Power is supplied to the emergency cooling chain only by the emergency power Diesel systems, i.e. even when there is no loss of off-site power.
- When the grid voltage returns after a case of loss of off-site power, the emergency power loads cannot be switched back without interruption.

The degree of redundancy of the uninterrupted emergency power supply is sufficient. On the other hand, the reliability of the uninterrupted power supply system is not up to requirements. The following are the most important shortcomings:

- Active transfer switching and control devices (thyristor switch, converter control) in the emergency power case.
- The reversible motor generators (RMG) are prone to breakdowns.
- There are no functionally separate rectifiers and power inverters.
- There is no double feeding of the DC buses or DC loads.
- The capacities of the batteries are too low. The discharge time of 30 minutes must be increased to 2-3 hours according to an RSK requirement.

- Physical Separation

The main items of equipment in the station service system, such as the 6 kV station service switching system and the 380 V main distributors, are installed in special rooms separate from the equipment of the emergency power systems.

The strands of the emergency power generation and distribution systems are physically separated. However, this physical separation is not maintained consistently in secondary cable routes and in the cable gallery underneath the main control room.

- Overvoltage Protection

When compared with the state of technology, the lightning protection and grounding systems show major deviations from national and international standards (such as DIN, IEC). The relays currently used, which are relatively insensitive to overvoltage, require backfitting only to a limited extent.

- Summary

The basic concept of the electric power supply is deemed to be suitable.

The following changes must be made:

- Improving the uninterruptible electric power supply.
- Increasing the battery capacities.
- Checking the emergency power balance and, if necessary, increasing the capacity of the Diesel generators.
- Improving the electrical systems for measurement, control and monitoring.
- Adapting the power supply system to the requirements resulting from the use of modern electrical installations and instrumentation and control systems.

6.3 Instrumentation and Control

In accidents and incidents, the instrumentation and control (I&C) systems important to safety have the following functions:

- Monitoring, limiting and shutting down the reactor power.

- Controlling the process engineering safety systems.

When reaching activation criteria, the control systems important to safety automatically initiate the required protective measures. The control systems important to safety are redundant and installed so as to be physically separate.

The I&C systems important to safety for controlling the emergency core cooling and residual heat removal chain (SAOS/GZ) is built in three trains, each train having two channels. The reactor protection system (SUS) has two trains and is based on the closed-circuit principle.

The conceptual design of the I&C systems important to safety is largely acceptable. One major exception is represented by the two-train design of the reactor protection system. During maintenance of parts of the reactor protection system, this would not allow to cope with a single failure.

In addition, some other requirements of nuclear codes and standards are not met:

- The following criteria for activating reactor scram are missing:
    - Activity in the main steam line high
    - DNB ratio low
    - Pressure in the primary system high
    - Pressurizer level high
- There is no diverse activation of reactor scram for the "pressurizer safety valve stuck open" accident
- Diverse detection of loss of off-site power requires the "frequency of emergency A.C. buses low" criterion
- No diverse signal processing because of the identical composition of measurement and control trains up to the initiating relay
- There is a lack of automatic limiters and protective actions, such as
    - fast cooldown of the unit via the secondary system,
    - ensuring sufficient shutdown reactivity during startup and operation

- A level probe in the reactor pressure vessel is missing

- The 30-minute rule (after the onset of an accident, no manual measures are required within 30 minutes) is not observed in the "steam generator tube leak" accident, among others

- The instrumentation is not qualified to work under accident conditions

- The accident monitoring instrumentation is insufficient

- Self-monitoring and automatic monitoring capabilities of the interlocks exist only in some areas

- In some areas, redundant systems are located in the same fire area

- Due to the layout of some instrumentation and control rooms, leaks in feedwater and main steam lines could damage technical instrumentation and control installations

- The ergonomic design of the main control room is poor

- Technical equipment is outdated

- Equipment is of low quality (no service-free equipment is installed in the engineered safeguards systems)

- There is no proof of component qualification

Because of conceptual weaknesses in the reactor protection system and the outdated and not very reliable control equipment, the entire instrumentation and control system would have to be exchanged to make the plant licensable.

6.4　Ergonomics

An ergonomically satisfactory design of the control rooms is essential to plant operation. Ergonomic defects can have a particularly negative impact on accident management.

In accordance with KTA 3904, workstations, working supplies, work sequences, and the working environment must be so designed as to create optimum preconditions for the workforce with respect to technical safety. Optimum functioning of the man-machine system must be ensured.

The following are factors of major influence upon the functioning of the main control room and the standby control room as a working system:

- Control room layout
- Layout of consoles and panels
- Legends
- Illuminated status indicators
- Alarms and reports
- Displays of quantitative information
- Controls
- Working environment
- Systems assisting in accident diagnosis
- Written support
- Education and training

In a technical working report, all these characteristics are analyzed and evaluated in detail. Recommendations are given for ergonomically better designs. As a yardstick, numerous rules and codes as well as the technical literature in the field are available (e.g. /1/ to /3/).

In summary, it must be stated that the control room of the unit as a working system exhibits a number of defects, from the point of view of ergonomics, and needs to be fundamentally redesigned. Only in this way can the conditions be created for optimum operator performance.

In Sec. 6.3, the replacement of all instrumentation and control systems is demanded. This automatically includes new control room designs. The control rooms must be rebuilt in the light of ergonomic aspects in such a way that the shortcomings encountered in Unit 5 are eliminated.

References, Section 6:

/1/ KTA 3904: Warte, Notsteuerstelle und örtliche Leitstände in Kernkraftwerken (Control Room, Remote Shutdown Station, and Local Control Units in Nuclear Power Plants).

/2/ KTA 3501: Reaktorschutzsystem und Überwachungseinrichtungen des Sicherheitssystems (Reactor Protection System and Monitoring Facilities of the Safety System).

/3/ KTA 1201: Anforderungen an das Betriebshandbuch (Requirements to be Met by the Operating Manual).

7 Spreading Impacts, Civil Engineering Aspects, Radiation Protection

7.1 Spreading Impacts

In planning the design of Unit 5, the possibility of hazards arising from spreading internal impacts was taken into account. All nuclear safety systems are designed with threefold redundancy (3 x 100%). The possible failure of one redundancy due to an initiating event (e.g. fire) and the simultaneous, independent failure of another system redundancy was included as a design basis.

A precondition for this concept to work is the reliable protection from common mode failures of the systems concerned.

For this purpose, design planning provided for

- large degree of physical separation of redundant safety-related systems,
- the installation of a standby control room,
- the definition of permissible repair times in the "Limits and Conditions of Safe Operation".

However, the requirement of independence of redundancy was not met completely in planning and building the plant.

7.1.1 Yardsticks Used for Reference

The laws, codes, and guidelines to be taken into account in the Federal Republic of Germany are used as a yardstick for reference in determining to what extent the design of Unit 5 of the Greifswald Nuclear Power Station is in agreement with the most important goals of protection set forth in those codes and standards.

Especially the following rules and regulations are considered in assessing spreading effects:

- Safety Criteria for Nuclear Power Plants (Criteria 2.6 and 2.7),

- RSK Guidelines for Pressurized Water Reactors, (especially Sections 11, 12, 18, 19),

- KTA Codes (e.g. KTA 2101, Fire Protection in Nuclear Power Plants; KTA 2202, Escape Routes in Nuclear Power Plants; KTA 2207, Protection of Nuclear Power Plants against Floods; KTA 2201, Designing Nuclear Power Plants against Seismic Impacts).

7.1.2 Spreading In-Plant Events

7.1.2.1 Fire

- Structural Fire Protection

Separation of redundancies by structural measures, with a fire resistance rating of F 90, is provided for important safety-related systems and components in the reactor building and the emergency power building. However, this separation is not maintained consistently at several points, especially in the reactor building. In some areas of that building, cables of two or three redundancies of important safety-related systems are run together. This contradicts the requirements in KTA 2101.1, e.g. in Sec. 3: "It must be ensured for redundant units of the safety system that a fire remains restricted to one such redundant unit." Subsequent partial coating with insulating layers was an attempt, in some cases, to protect the cables from the consequences of fires at least for some length of time. In this respect, each individual case will have to be examined for the extent to which existing fire protection measures, in the light of the fire loads existing in those areas and also of the relevance to safety of the cables run here, provide sufficient preservation of their functioning, and what additional measures (such as fire walls, coatings, installation of extinguishers) may have to be taken. If the functioning of safety-related cables is preserved sufficiently well, the goal of protection as specified in the above KTA code may be considered to be met.

Whether the cable layout meets the design requirements or the yardsticks currently applied in evaluation needs to be examined in every detail in an extensive, complicated spot check. Cable layouts not yet meeting present-day requirements must be backfitted (e.g. by the use of fire walls, coatings, extinguishers).

The quality of cable compartments, insulating cable coatings, and fire protection doors is inadequate in some cases; again, major backfitting measures are necessary.

In the ventilation systems, fire protection dampers exist at some spots only. The concept of protecting the ventilation systems is based on the switch-off of the ventilation systems in cases of fire. However, this alone will not exclude the possibility of a fire and the fumes respectively spreading through the ventilation system in certain areas. In KTA 2101.1, a requirement in Sec. 4.4 reads that "ventilation systems in compartments and areas with redundant units of the safety system must be arranged and built in such a way that a fire in one redundant unit will not affect the functioning of the other redundant units by way of the ventilation systems." Fire protection dampers in areas with safety-related divisions and in staircases (protected escape routes) must be backfitted.

Also for reasons of fire protection, the main control room and the standby control room must be isolated (systems coupling, cable routing).

The entire turbine hall for the eight power plant units constitutes one fire area. As a consequence of the tremendous fire loads existing in the turbine region (e.g. main oil tank with 56 m³ of oil per unit; elevated oil tank; cable insulation), and also because of the presence of ignition sources in the turbine building, large area fires cannot be excluded. As the turbine hall also contains installations related to safety (such as the feedwater and emergency feedwater supply systems), a large area fire could cause several redundancies to fail in units not separated with respect to fire protection. This contradicts the requirements in KTA 2101.1, among others.

In order to achieve the goals of protection as set forth in the codes and standards about fire protection, at least a fire resistant encapsulation of the individual redundancies of engineered safeguards systems and components is necessary. From the point of view of fire protection, it would be more effective to move safety-related systems and components from the turbine hall into a separate building and to separate redundancies in the light of aspects of fire protection.

Irrespective of the technical safety requirements with regard to fire protection in the turbine hall, measures must also be taken under conventional building codes to avoid large area fires from starting and spreading (e.g., encapsulation of the main fire loads) so that the necessary special permits can be obtained for oversized fire areas and lengths of escape routes respectively.

Irrespective of safety-related problems, conventional criteria under the building codes covering preventive fire protection are also violated at numerous points in the plant. This applies, in particular, to securing the escape routes (e.g. free passage, consistent structural separation of staircases, keeping corridors free from fire loads), to the length of escape routes (in some rooms of the equipment building and, in particular, in the turbine hall), and to the permissible sizes of fire areas (in the turbine hall). In each of these cases, systematic examination is required.

- Technical Fire Protection

The safety-related areas of the plant (the turbine hall only partly) are monitored by automatic fire detectors. Within the confinement system, a modern fire detection system with addressable sensors and computer-aided signal processing capability has been installed. In addition, the area of the main coolant pumps is monitored by a video camera system. Fires are indicated simultaneously in the control room and to the plant fire brigade.

As the fire detection system is made up of a number of systems, not all of which are perfectly harmonized, the whole concept and the detection system need to be verified. Upgrading is necessary in some areas.

With the exception of the area of the main coolant pumps, water sprinkler extinguishing systems have been installed or planned in all areas containing major fire loads. The water sprinkler extinguishing systems are actuated manually, except for the area of the elevated oil tanks and the unit transformers. As the extinguishing systems are not designed in line with requirements and as the reliability of manual startup is low, the whole concept needs to be revised and, if necessary, the water sprinkler extinguishing systems must be upgraded.

In the area of the main coolant pumps, where oil leaks are bound to occur as a result of plant design, halon extinguishers have been installed. According to the present state of knowledge, they cannot be used to protect rooms during oil fires. Effective, quick-acting, stationary extinguishing systems must be installed instead.

The characteristics of the extinguishing water pumps (0.95-0.98 MPa pumping pressure, maximum capacity 510 m³/h) are sufficient. The pumps can be connected both to the emergency power system and to the grid system of the neighbouring generating unit. The pumping capacity can be further increased by adding the pumping capacities of the fire tank trucks of the plant fire brigade.

Water for fire fighting in Units 5 and 6 is supplied through a common ring piping system from the fire fighting water system of Units 1 to 4, from the waterworks, and from an additional pumping station at the intake canal, which thus can be considered to be reliable.

In case of fighting a fire, redundant systems outside the fire area must not be impaired by the water. Cable ducts normally are equipped with devices for collecting and removing water from fire fighting. Whether this criterion is met in the plant needs to be studied systematically.

- Operational Fire Protection Measures

The concept of fire fighting in the plant is based mainly on the plant fire brigade, relying on its quick alert capability.

- Summary Evaluation and Recommendations

(1) The deficiencies in fire protection, as far as the reactor building is concerned, result mainly in the combination of cable routes of various redundancies of systems with important safety-related functions. The three redundancies must be separated in fire protection.

(2) The quality of fire breaks, insulating cable coatings, and fire protection doors is partly unsatisfactory; backfitting measures are necessary.

(3) Fire protection dampers exist in the ventilation systems only at a few points. Backfitting is required in areas with separations required for safety-related reasons and in staircases.

(4) As the fire alarm system is made up of various subsystems of which not all are properly harmonized, the basic concept and the sensors need to be checked.

(5) Oil leaks occur in the area of the main coolant pumps. The halon extinguishers used in this area are not able to protect compartments. Effective extinguishers must be installed.

(6) The fire loads and ignition sources existing in the turbine hall do not allow large area fires to be excluded. Safety-related systems should be protected by being moved into a separate building.

(7) For reasons of conventional building codes, measures must be taken in the turbine hall to prevent the generation and spreading of large area fires (e.g. by encapsulation of the main fire loads) so that the necessary special permits can be granted with respect to excessive sizes of fire areas and lengths of escape routes. In addition, systematic checks, especially of the escape routes, must also be carried out in other areas of the plant.

(8) The functions of the main control room and the standby control room are not isolated completely. Such isolation is necessary for reasons of fire protection.

7.1.2.2 Flooding

- Intake Structure and Pump Building for Service Water A

The intake structure and pump building contain the three redundancies of the service water system A for each of the reactor Units 5 and 6. The two pump systems making up one redundancy are installed in separate chambers. These pump chambers have no top covers, and the pressure-side pipe penetrations through the chamber walls are not sealed.

Pipe failure, e.g. as a result of corrosion and erosion damage to the pipe ducts adjoining the pump chambers, may impair the functioning, up to the point of failure, of all legs of the service water system A through the pressure-side pipe penetrations and, in case of the water level continuing to rise, through the building corridor at +3,85 m. Structural measures must be taken to prevent reliably flooding of the pump chambers through the pipe penetrations or the building corridor.

As another precautionary measure, a reliable system for leakage detection must be installed in the pump chambers and pipe ducts, and the alarm level setting must be incorporated as a safety-related measured quantity in the "limits and conditions of safe operation".

- Turbine Hall

The drives of the pressure-side valves of the emergency feedwater system are installed at approx. 1 m above the lowest level (-4.5 m) of the turbine hall. One emergency feedwater pump and the five main feedwater pumps are located at -2.1 m. The other two emergency feedwater pumps are located at ±0.0 m (ground level).

Failure of a main cooling water line, e.g. as a consequence of dropping loads, without timely shutoff of the pump will cause flooding of the emergency feedwater valves in the areas separated by partitioning walls in Units 5 and 6, and of the pumps located at -2.1 m. For this case, the partitioning brick walls are supposed to withstand the water pressure.

The components of the emergency feedwater system must be arranged and located so that there can be no flooding hazard and that the principles of physical separation of redundant subsystems are observed; failing this, other substitute measures must be taken. Due to the large flooding capacity of the turbine hall areas pipe failure in other systems, such as the service water systems A and C, will give rise to hazards only in a matter of several hours, thus leaving enough time for the situation to be detected and countermeasures to be initiated.

- Reactor Building

Three physically separate chambers at the lowest level (-3.6 m) contain the subsystems of the HP emergency cooling system, the LP emergency cooling system, and the sprinkler system arranged so that the subsystems making up one redundancy are co-located. The chamber doors are fitted with seals and are locked from the outside. Drains provided in the chambers are equipped with sealing systems which can be locked.

In case of failure of the service water system A or the boric acid storage tank in one chamber, flooding of other chambers must be prevented. The partitioning walls of the chambers and the installed packings of the penetrations must be demonstrated to withstand the water loads arising in case of flooding. Moreover, water must be prevented from flowing over into adjacent chambers through the drainage system. For this purpose, the suitability of the shutoff systems installed must be demonstrated. In normal operation, the shutoff systems must be protected in the closed position. In order to avoid inadmissible consequences of failures and of maloperation respectively of the filling system for the boric acid storage tanks, the valves installed in the pipes spreading beyond redundancy areas must be protected in the closed position in normal operation.

Flooding of a chamber and the loss of one redundancy of the emergency cooling systems can be controlled, and the safety-related conditions of down times and repair times respectively of the affected systems are met.

To reduce the rate of flooding events, a qualified and reliable system for leakage detection must be installed which allows the shift personnel to take effective countermeasures.

At the lowest level of the reactor building, there are the two pump units for the cooldown system of the spent fuel pit. Failure of a service water pipe may cause flooding of the pump systems and complete failure of the pit cooling system. The loss of both redundancies of the pit cooling system may give rise to inadmissible temperature rises in the spent fuel pit. Flooding of the pump units therefore should be prevented by precautionary measures.

Cold water leakages in the steam generator box (leakages of systems running hot are treated in Sec. 5.1) give rise to water inrushes in the compartments below the steam generators, main coolant lines, and pumps. As the steam generator box has a high flooding capacity, cold water leakage will not impair the engineered safeguards.

As a consequence of the location and physical separation of the Diesel generators, no inadmissible consequences are expected to arise from cooling water leakages.

- Summary Evaluation and Recommendations

(1)   To avoid sequences of events spreading beyond the redundancies as a consequence of major leakages in the main cooling water system, the service water system, and of emergency cooling water, structural evidence or measures are required to be able to eliminate the deficits resulting from the

lack or insufficient demonstration of physical separation. Cross connections spreading beyond redundancies through drainage systems or pipes of the filling system must be secured in the closed position during normal operation. This applies to the following building areas with important safety-related systems:

- Intake structure for the service water system A:
service water pump units.

- Turbine hall:
emergency feedwater pumps and feedwater pumps.

- Reactor building:
HP and LP emergency cooling systems and sprinkler systems,
pit cooling system

(2) To reduce the probability of occurrence of inadmissible flooding events, provisions must be made to ensure qualified leak detection and avoid maloperation.

7.1.2.3 Other In-Plant Spreading Impacts

Other in-plant impacts, such as missiles arising from turbine failure and blast waves produced by rupturing vessels, were not taken into account in the design planning of Unit 5. No sufficient documents are at present available to assess the potential consequences and the necessary upgrading measures.

If safety-related systems were moved from the turbine hall into a separate building, any consequences arising within the turbine hall would be of secondary importance only. In that case, mainly the impacts on installations within the reactor building would have to be taken into account.

7.1.3 External Impacts

The power plant has not been designed against external impacts, such as airplane crashes, earthquakes, and explosion blast waves. To what extent it has been designed against floods cannot be determined at the present time. In addition, no data are available to show the probabilities of flood water levels occurring.

In assessing protective measures against external impacts a fundamental distinction must be made between events which

- must be considered as design basis accidents, such as earthquakes, floods, wind and snow, and similar events, and those

- for which measures must be taken to minimize the residual risk (airplane crash, external blast waves from chemical reactions, external impacts of hazardous substances).

Provisions against design basis accidents must be made in any case. Wind and snow loads normally are accommodated in conventional building design. The maximum possible structural loads are determined within the framework of structural investigations. These investigations also indicate to what extent protection is afforded against seismic events. In designing against floods, the criteria listed in KTA 2207 (Protecting Nuclear Power Plants against Floods) must be observed. The locations and layout of buildings do not indicate any flood hazards existing for the reactor building and the turbine hall. To what extent there is danger of the intake structure and the pump building being flooded can only be determined once the probabilities of occurrence of floods are known. If necessary, specific protective measures would then have to be taken.

Events resulting from an airplane crash, from external blast waves caused by chemical reactions, and from external impacts of hazardous substances are no design basis accidents in the sense of Sec. 28, Subsec. 3 of the German Radiation Protection Ordinance. Any measures taken against such events serve for risk minimization and, consequently, for the protection of the public. Such measures are taken in accordance with the Safety Criteria for Nuclear Power Plants, the RSK Guidelines, and the Guideline for the Protection of Nuclear Power Plants against Blast Waves from Chemical Reactions. In the light of site-specific conditions and probabilistic considerations, it would have to be investigated whether and to what extent such impacts are of importance.

7.2 Civil Engineering Aspects

Investigations were made to find out whether the buildings and other structural facilities falling under the license under the Atomic Energy Act correspond to the structural engineering design criteria required in accordance with the state of the art, and what deficits exist, if any.

- Building Structure

In these investigations, the reactor building, the floor building with the ventilators and the exhaust air stack, the central building, the turbine hall, the floor building containing the control rooms, the special building with the connecting bridges, the emergency power system, and the intake structures for the main cooling water and service water systems were examined. Attention was paid in particular to the reactor building. Work was restricted to the main load bearing structures and the main impacts.

Most buildings, including the reactor building, were planned in detail under the responsibility of the Soviet Union. For projects not planned in the GDR, the GDR had waived the right to reexamine the stress analysis and the building design. Only the working plans were made available to the GDR. No complete and testable stress analyses are available for examination.

The reactor building consists of a hermetic and a non-hermetic part. While the latter is a conventional, partly solid, partly skeleton-type, structure made of reinforced concrete, the walls of the confinement (hermetic part) were made as a composite steel cell structure. This structure is made up of prefabricated steel cells welded together and filled with concrete on the construction site. The steel cells consist of external metal sheets, normally 6 mm thick, which act both as static reinforcements and as sealing liners. At a few highly loaded points, the sheet metal reinforcement was supplemented by round bar steel. This is found especially in corners, where the reinforcement was designed in an almost conventional way and then welded, e.g. to the sheet metal or to the anchorage devices of the sheet metal. In those areas, the sheet metal almost exclusively has sealing functions.

The composite steel cell construction technique was developed in the GDR under the leadership of the Bauakademie (Construction Academy). The results of this development, including a complete set of codes about the composite steel cell type of construction, were delivered to the Soviet Union where they were modified slightly and taken into account in planning the construction of the plant.

Assessing the design of the building structures of the main building was rendered difficult by the absence of tested or testable stress analyses, complete sets of blueprints, and any indication by the Soviet Union as to the load assumptions and load combinations employed in designing the structures. For the offhand calculations performed, the dead weight loads were determined roughly, and the loads arising from the equipment were estimated.

- Slant of the Reactor Pressure Vessel

Settlement calculations were performed for the problems of settlement and the slant of the reactor pressure vessel in such a way that the sequence of building steps was taken into account. According to present knowledge /1/, an inclination of the reactor pressure vessel of $\Delta\varpi = 1.3$ to 1.4 mm is expected to result from the settlement of the building structures. A slant of $\Delta\varpi \geq 1.5$ mm is considered to be most unlikely. The calculations also have shown the bottom slab to be overloaded in the area of the partition joint (axis 22). However, this does not jeopardize stability, although problems could arise with respect to building insulation.

- Loads Arising from Loss-of-Coolant Accidents

The investigation was carried out for the maximum load, namely the double-ended break of a main coolant line of NB 500. A maximum internal overpressure of 0.15 MPa at a temperature of 127 °C was assumed. Analyses of the load bearing capacity were conducted on the basis of simplified models. The loads arising in this load case will be accommodated with a very high degree of probability.

- External Impacts

The earthquake, airplane crash, and explosion blast wave load cases were considered. The building structures were not explicitly dimensioned to withstand earthquakes. A comment on seismicity /2/ states for Western Pomerania that intensity level IV on the 12-level MSK-64 scale has never been exceeded so far. The maximum anticipated ground accelerations are indicated as $a_h = 0.20$ m/s², which would correspond to intensity level V. Consequently, the seismic intensity on the Greifswald site is so low as to require no special design criteria.

Designing new plants and assessing existing ones can be based on the minimum design criteria specified in KTA-2201.1. Accordingly, a minimum level of $a_h = 0.5$ m/s² must be assumed for the design basis earthquake, irrespective of the maximum acceleration to be expected on-site. As the calculations performed have shown, global structural failure of the main building as a consequence of such an earthquake is not to be expected. However, some partial structural backfitting measures are required which appear to be feasible. The floor acceleration responses, e.g. at +14.5 m in the reactor building, are on the order of 4.5 m/s² at frequencies of 1-4 Hz.

The airplane crash load case cannot be accommodated by the building structures, the explosion blast wave only within certain limits. Upgrading measures aiming at "full protection" are hardly feasible.

- Turbine Explosion

As a result of the turbine arrangement in the turbine hall parallel to the reactor building, a turbine explosion entails the danger of shooting missiles at the reactor building; however, this risk can be reduced by structural protective measures.

- Summary

All buildings, except for the reactor building, have the character of conventional industrial buildings and are designed accordingly. The airplane crash and explosion blast wave external impacts cannot be accommodated.

As far as building construction is concerned, a permit under present codes could be granted. However, this would require a special approach to be used for the load cases of airplane crash and explosion pressure wave.

7.3 Radiation Protection

7.3.1 External Impacts of Normal Plant Operation

The potential radiation exposure of the public resulting from discharges of radioactive substances with the exhaust air and the liquid effluent in the normal operation of Unit 5 was determined under the assumption that this unit discharges one eighth annually of the maximum permissible discharges defined for this site. On the whole, the radiation exposures determined show that the dose limits under Sec. 45 of the Radiation Protection Ordinance for discharges of radioactive substances can be observed by Unit 5 on the Greifswald site. The operation of additional nuclear generating units on the same site requires further considerations to be made in limiting the discharges of radioactive substances.

7.3.2 Radiological Protection of Workers

Due to the fact that the materials used in Unit 5 of the Greifswald Nuclear Power Station contain only small amounts of cobalt, only a small accumulation of gamma emitters in the plant is to be expected. In preparing maintainance work, the operator in addition conducted intensive studies of decontamination techniques which were applied already in Units 1 to 4. In the light of these conditions, relatively low exposure levels of two to three person-sievert mean collective dose per annum were achieved for Units 1 to 4, which is a quite good value for a unit in operation for a longer period of time.

However, the studies of Unit 5 indicate that additional measures can be taken to reduce the radiation exposure of the personnel in Unit 5. This is necessary as aspects of radiological protection of workers were not taken into account sufficiently in the design of the plant, and as the far less favorable maintenance and repair conditions in a number of plant areas, compared to Units 1-4, make for less satisfactory conditions with respect to the radiological protection of workers. In comparable maintenance and repair work in Unit 5, therefore, the radiation exposure of the personnel is expected to be higher than in the older plants. The reasons include, in particular, the worse accessibility of components and systems, the larger number of systems in the steam generator box, with space problems resulting, and impossibility of optimal use of lifting gear in the sense of radiation protection. All these conditions result in shorter distances from the radiation sources and longer residence times for work in radioactive areas. The operating experience gained from Units 1-4 shows that possible measures to reduce the exposure were not taken into account.

On the whole, radiation protection of the plant personnel is seen not to be in line with present practice in the nuclear power stations in the Federal Republic of Germany, and to require improvements in a number of points so as to meet the minimization requirement in the Radiation Protection Ordinance in accordance with the Guidelines for the Radiation Protection of the Personnel during Maintainance in Nuclear Power Plants with Light Water Reactors.

In the underlying studies, a number of specific problems of radiological protection of workers were determined. The implementation of the following steps is thought to be required to correct these deficiencies:

(1) The present in-plant radiation protection regulations need to be revised so as to meet the criteria of the Radiation Protection Ordinance and the applicable codes and regulations. Moreover, the organization, competences, and duties of radiation protection must be revised in the light of an appropriate quality assurance system in order to eliminate the deficits now obviously existing in organization.

(2) The use of a suitable, directly readable personnel dose monitoring system for occupationally exposed personnel is necessary. In addition to dose monitoring, the system must also have capabilities of dose warning, monitoring of the access to the control area, and electronic data evaluation and processing. Moreover, radiation protection surveillance of accessible rooms and areas must be improved.

(3) Radiation exposure must be reduced by increased mechanization and the use of remote handling techniques, especially for work on the entire primary system, including the reactor pressure vessel, the pressurizer and steam generator, and, if necessary, for testing and surveillance work in the steam generator box. In this respect, the additional detailed measures already identified by the operator in the Units 1-4 to reduce radiation exposures must also be implemented.

(4) Modifications in building structures, such as moving the staircase near the lock platform (Room G202A) into an area with a lower local dose rate, and modifications to the doors as well as changes to the lock systems of the transfer lock and the emergency locks respectively must be made.

(5) The use of high-quality respiration protection gear.

References, Section 7:

/1/   Comment by Prof. Nendza, Erdbaulaboratorium Essen.

/2/   Comment by Prof. Schneider, University of Stuttgart.

8   Evaluation of Operating Experience

8.1   Work Performed

In the engineering assessment of the plant concept and of the safety-related design of Unit 5, the operating experience existing to date was also evaluated.

This evaluation serves to determine whether

- the frequencies and types of events,
- the sequences of those events,
- the frequencies of component and systems failures

provide any information about

- design deficiencies in the combined action of systems functions,
- defects in the design of systems and components,
- deficiencies in component reliability during operation and challenge,
- shortcomings in plant management.

Abnormal events in Unit 5 were reported and detected in accordance with Directive 1/88 of the State Office for Atomic Safety and Radiation Protection (SAAS) on notifiable events. Notification to SAAS became mandatory after the commissioning permit had been granted on December 30, 1988. Moreover, events below the notification limit, which herein are called unplanned events (UE), were also collected and evaluated. Unplanned events were notified on the basis of an in-house directive by the plant vendor about the detection and handling of unplanned events during the commissioning of Unit 5 (Annex 8 to the Commissioning and Plant Regulations (IAO), Nord III/IV, of November, 10, 1988).

The evaluation is based on 365 events notified until September 30, 1990. Of these, 164 events (unplanned events) are below the notification limit, while 189 events (abnormal events-3) are of little significance to engineered safeguards, and 12 (abnormal events-2) are of greater significance with respect to engineered safeguards.

No events of the highest level (abnormal-1) were notified.

In trial operation of the plant so far, the reactor was critical for 2693 h (112 d) at a power not exceeding 55% of the design power; on 66 days of that period, it was synchronized to the grid.

As major loads acting on the plant a total of 20 reactor scrams was observed. In addition, there were three events where sizable amounts, and three other cases where very small quantities of cold water were fed into the primary system. In the secondary system, there were four cold water feeding events through the emergency feedwater nozzles in two steam generators, and two cold water feeding events in two other steam generators. These were due to incorrect activation.

When the events were analyzed, they were subdivided systematically into categories of events. For events affecting process systems, a distinction was made between defects in the control and protection system of the reactor, defects in the primary system, breakdowns and defects in the emergency cooling system, failures in the feedwater system, leakages in the primary and secondary systems, and defects in the pressure protection of the pressurizer and the steam generators. Events affecting electrical and instrumentation and control systems etc. were subdivided into defects of the emergency power Diesel systems, failures of the power supplies to safety-related loads, and failures in instrumentation and control. Events indicating the existence of administrative shortcomings were subdivided into violations of the "Limits and Conditions of Safe Operation" and failures due to insufficient plant documentation. The safety-related events not included in this classification were summarized in a class of their own.

In the defects which occurred, a distinction must be made between typical commissioning failures and characteristic technical defects and systems failures of Unit 5.

The weak spots identified in the analyses of individual events (see technical working reports) give rise to the following major requirements in terms of upgrading measures, the implementation of which is a prerequisite for licensability being achieved.

8.2 Upgrading Measures Required

8.2.1 Mechanical Systems

(1) Development and application of suitable materials testing techniques for quality assurance of passive mechanical components installed.

(2) Verification of the secondary system's sufficient protection against entry of sea water (especially in the area of shutdown condensers).

(3) Enhancing the reliability of and, if necessary, exchanging all components of the atmospheric main steam dump station (BRU-A) and the bypass stations, the valves important for pressure protection of the steam generators, and the valves in the feedwater and condensate systems.

(4) Decoupling the entire nitrogen supply system by means of check valves, and monitoring by differential pressure measurements.

(5) Interlocking the crane control system in such a way as to prevent collisions with plant components as a result of incorrect operation.

(6) Construction and installation of the rupture disk in the pressurizer relief tank so that it will respond reliably only after the actuation pressure has been reached.

(7) Fitting the Diesel startup air system with installations for air drying and for dewatering compressed-air tanks and pipes respectively; making compressed-air tanks and pipes of corrosion-resisting steel.

(8) Ensuring smooth movement of the control linkage actuating the fuel injection system of the emergency Diesel.

(9) Fundamental revision of main coolant pumps and its oil systems in order to eliminate oil leakages.

(10) Checking the stud bolts of the main coolant pumps when replacing a gasket (materials inspection), and also checking the bearing connections and throttles; if necessary, replacement by a new design.

8.2.2 Instrumentation and Control

The instrumentation and control concept must be revised and established in a fault-tolerant technology, the suitability of which has been verified in advance. Also, testability and automatic failure monitoring capability must be improved. The single-failure criterion must be taken into account in these steps. The instrumentation and control concept must incorporate measurement and control systems, limiter and protection systems, and the signaling concept. The following are the measures to be taken and implemented respectively:

(1) Checking the completeness of the reactor protection activation criteria.

(2) Using measuring systems resistant to accidents.

(3) Automatic comparison of the readings of multichannel measurements, and automatic signaling in the case of inadmissible deviations; avoidance of different signal statuses in the control room and the interlocking logic.

(4) Improving signalization in the control room for accidents and wrong setting of valves in safety systems (monitoring operational position).

(5) Exchanging all relays used in safety-related circuits for instrumentation and control units of proven suitability.

(6) Verifying the reliability of electric contacts, e.g. for the control and protection drives.

(7) Improving the reliability of computers for logging the main systems parameters in the course of disturbances; if necessary, replacing existing computers.

(8) Checking the power supply concept for safety-related measurements and interlocks, including their electrical protection systems, on the basis of a failure effect analysis in order to prevent failures of protective actions or

actuations of protective actions whose impact on safety is not evident, and also to avoid other erroneous actuations, e.g. demeshing of the safety systems.

(9)   Using automated test circuits.

(10) Prohibiting bridging automatic fuses when running checks; modifying the instrumentation and control concept in such a way that test circuits can be established without changing the wiring.

(11) Replacing, by an automatic control system, the manual bridging of the "last turboset trip" criterion.

(12) Improving the activation of valve drives.

(13) Using function group control systems to avoid incorrect operation (e.g. feedwater supply, condensate pumping, main coolant pump).

(14) Revising the Diesel control system so that defects due to overloading are prevented under any operating condition (also in cases of maloperation).

(15) Automating the neutron flux measurement system so as to avoid incorrect operation.

(16) Defining limit conditioning for measurements of reactor power.

(17) Regulation of the control room ventilation system so that single failures will not give rise to an inadmissible operating status (improvement in signaling for control room monitoring).

(18) Reliable monitoring of the operating condition of the main coolant pump by suitable criteria, such as rotational speed.

(19) Reliable monitoring of safety-related filling levels, e.g. pressurizer, steam generator, accumulator.

(20) Automatic monitoring of sumps in the reactor building with signaling to the control room (hazard reporting system).

(21) Improving leak detection.

(22) Improving the quality of the boric acid concentration measurements.

8.2.3 Station Service Power Supply

(1) Designing the station service power supply system and the electrical protection system in such a way that defects in transformers, switching systems, distribution networks and loads can be detected and shut off reliably.

(2) Verifiable calculation of the maximum and minimum short circuit currents and determination of the minimum voltage limits for all loads.

(3) Diode-decoupled double feeding on the DC-side of safety-related loads; use of two batteries per train.

(4) Evidence of sufficient reliability of installations for uninterrupted power supply; separation of the functions for battery charging and for supplying the safe main distribution board.

(5) Demonstrating the suitability of all switching systems of safety-related loads at all voltage levels; where necessary, exchanging switching systems for those whose suitability has been verified.

(6) Verifying the suitability of the cables in the station service power supply system.

8.2.4 Building Structures

(1) Testing and eliminating, if necessary, the fire hazard on the 14.7 m platform.

(2) Ensuring and demonstrating sufficient separation of the Units 5 and 6 in such a way that inadmissible mutual interference (e.g. through sumps of the common special building) is avoided.

(3) Evidence of safe separation of the overflows from the building and the leakage water sumps.

8.2.5 Plant Organization, Operating Instructions, Quality Assurance

(1) Bringing plant documentation up to the actual plant design level.

(2) Establishing a reliable central modification service for the plant documentation.

(3) Compiling an operating manual in line with the requirements contained in the operating manuals of nuclear power stations in the Federal Republic of Germany.

(4) Evidence of sufficient qualification of the personnel.

(5) Improving the job order and isolation concept for better coordination of the activities of shift personnel, radiation protection, fire protection, maintenance, and technical departments.

(6) Clear definition of competences of the architect engineer and the power plant personnel.

(7) Establishing a revision concept to ensure nuclear safety.

(8) Defining the nominal set points of valves and switches for various operating states; if necessary, adapting the technical design.

(9) Intensifying quality assurance, from manufacturing up to the installation of components; demonstrating sufficient functioning capability and the quality required for all safety-related components.

(10) Comprehensive tests of the interaction of systems in the non-nuclear commissioning phase; reliable removal of temporary commissioning setups.

(11) Revising the concept of feedwater control from the LP preheaters up to the steam generators; successful completion of systems testing of the feedwater system prior to recommissioning of the reactor.

(12) Regular checking of measuring stations at sufficiently short intervals and regular checking of the drainage sumps by visual inspection.

(13) Forbidding crane operation over safety-related systems while the plant is run in power operation if no technical measures of protection are possible.

(14) Revising the mode of operation of the laboratory (rules for performing analyses; organization).

8.3 Summary Assessment

The specific weaknesses of the equipment correspond to the operating experience accumulated from Units 1-4.

In addition, Unit 5 revealed problems in the area of circuit breakers not known in Units 1-4.

The deficiencies in the equipment are due to weak spots in components resulting from manufacture, but mainly also due to the long periods of construction and commissioning. Counteracting these findings by additional testing and maintenance does not seem to be very promising. In addition, this strategy would entail the risk of creating negative impacts on the engineered safeguards features of the plant then in operation. This hazard is apparent from an event on April 21, 1990, when cleaning the contacts in the relay panels caused the scram valves to be actuated and shut off inadvertently.

In summary, the operating experience in the present state of the plant reveals major weak spots in the design and considerable weaknesses in the construction of systems and components. Operating experience so far has not indicated the existence of weak spots which could not be repaired in principle. Eliminating the weak spots which have been recognized requires extensive measures to be taken as a prerequisite of plant licensability.

9 Summary

Safety investigations of Unit 5 of the Greifswald Nuclear Power Station were carried out on behalf of the German Federal Ministry for the Environment, Nature Conservation, and Reactor Safety (BMU). For this purpose, the safety-related design of the plant was assessed, and the experience accumulated in commissioning the plant was evaluated. The extent was examined to which the plant meets the safety-related codes and regulations valid in the Federal Republic of Germany. Where valid codes and regulations were not met, investigations were conducted to see

- whether such deviations gave rise to a deficit in terms of safety,

- what measures could be taken to make up for or eliminate such safety deficiencies.

The design, in terms of reactor physics and thermal engineering, of the advanced type WWER-440/W-213 reactor largely corresponds to the design of the older type WWER-440/W-230. Irrespective of the different design features of the two lines, the nuclear power plants of the type WWER-440 have safety-related properties which must be judged positively:

- Low power density of the reactor core.

- Large water volumes in the primary system and on the secondary side of the steam generators,

- Shutoff capability of the main coolant lines.

Compared to Units 1-4 (W-230), Unit 5 (W-213) of the Greifswald Nuclear Power Station is equipped with considerably improved engineered safeguards. Thus, the safety systems in Unit 5 have higher capacities and, for the most part, are designed redundant as 3 x 100% systems. To a high degree, they are performed separately from the operating systems.

Like other plants of the W-213 line, Unit 5 has an emergency core cooling system and a residual heat removal system, the designs of which were based on the entire spectrum of possible leakage accidents up to the double-ended break of a main coolant line.

The plants of the type W-213 have a confinement system with a pool-type pressure suppression system to be actuated in loss-of-coolant accidents. Also, this system has been designed against the double-ended break of a main coolant line.

On the other hand, deficiencies apparent in Units 1-4 (W-230) have not been eliminated in Unit 5 (W-213). This is true, in particular, of

- the physical concentration of all main steam and feedwater pipes on one platform in the turbine hall (14.7 m platform);

- direct cooling with sea water of the heat exchangers of the emergency core cooling system and the residual heat removal system as well as other important safety-related loads;

- insufficient fire protection measures;

- failure to take into account loads arising from such external impacts as airplane crashes and explosion blast waves;

- unfavorable arrangement of the turbines relative to the reactor building (possibility of consequential damage arising from turbine failures).

In most of the examined cases, upgrading measures have been suggested to eliminate existing deficits. In some cases, further studies are necessary before a decision can be made about the upgrading measures necessary and possible respectively. However, even after upgrading of the plant there will still be deviations from the requirements under safety-related codes and regulations.

The plant has no containment in the sense of the term as it is used in the Safety Criteria of the Federal Ministry of the Interior (BMI), as the confinement is not protected by another enclosure. For this reason, there is no possibility of complete and controlled leakage extraction. Despite these differences relative to valid safety-related codes, the assessments conducted so far seem to indicate that the accident planning levels under Sec. 28, Subsec. 3 of the Radiation Protection Ordinance will be observed in design basis accidents and their radiological consequences.

Loads arising from an airplane crash cannot be accommodated by the building structures because complete protection by such building structures is almost impossible to achieve.

According to valid regulations, nuclear power plants must be protected against airplane crashes by building construction measures designed to minimize risk. An airplane crash is not a design basis accident in the sense of the Accident Guidelines. It still needs to be investigated whether observance of the requirements of risk minimization as defined in the RSK Guidelines is indispensable on the site of Unit 5, or whether administrative measures could be taken to reduce the risk to a sufficiently low level.

The assessment of pressurized components did not reveal shortcomings which could not be eliminated in principle. For a final evaluation of the components, the quality documentation kept by the manufacturers will have to be inspected in addition.

Although the effectiveness of emergency cooling so far has been analyzed only in some points, engineering assessment seems to show the emergency cooling systems to be sufficient in design. Calculations were performed of the effectiveness of the confinement system with pool-type pressure suppression. The models of the pressure suppression processes used in those calculations still need to be verified experimentally.

The backfitting measures derived from the examinations and investigations so far and the documents and analyses required for further tests are listed completely in Annex A.3. The following are important upgrading measures:

- Replacement of the instrumentation and control and redesign of the control rooms.

- Upgrading of the station service power supply system (switching systems, uninterrupted power supply, etc.).

- Construction of a self-sufficient emergency standby system consisting of a steam generator emergency feedwater supply system, an additional boration system for diverse reactor shutdown, a reactor protection system, and an emergency control room.

- Construction of a redundant nuclear component cooling system for items to be cooled with important safety-related functions.

- Protection against spreading impacts (consequential failures, fires, turbine explosion) of the pipes and equipment of the main steam and feedwater systems on the 14.7 m platform in the turbine hall.

- Upgrading of the pressure protection installations in the primary and secondary systems.

- Rerouting of the small pipes in the confinement.

In planning and preparing all these measures, detailed investigations must be conducted to see whether such measures might not be associated with impairments of technical safety in the whole plant.

If the proposed upgrading measures are carried out, no decisive conceptual defects will be apparent which, at the present status of investigations, would fundamentally jeopardize the plant commissioning and power operation from the technical point of view. Further analyses and verifications are still needed for a final evaluation to be made. This is true, in particular, of materials investigations and accident analyses for design basis accidents.

For further evaluation of the plant, the work performed so far should be supplemented by a detailed safety analysis in which probabilistic methods will also be used. For this purpose, not only the operating experience accumulated in commissioning Unit 5 but also the operating experience from other plants of the type W-213 should be evaluated. Such an analysis will also allow a quantitative check to be made of the sufficiency of systems reliability and the balanced character of the safety-related design.

10 Comment by the USSR Ministry for Atomic Energy Industry on the Safety Assessment of Unit 5 of the Greifswald Nuclear Power Station

10.1 Introduction

This comment refers to the GRS report on the safety assessment of Unit 5 of the Greifswald Nuclear Power Station and can only be transferred to the Units 6-8 of the Greifswald Nuclear Power Plant. These units differ from the other units of the type W-213, which are conformly designed.

The organizations of the Ministry for Atomic Energy Industry of the Soviet Union participated in drafting the comment, among them

- the Kurchatov Institute for Atomic Energy as scientific leader,

- the OKB Gidropress design organization as main designer of the reactor plant,

- the Energoprojekt Institute in Moscow as project engineer of Unit 5 of the Greifswald Nuclear Power Station.

The comment contains statements by the scientific leader, the main designer, and the general project engineer of the Greifswald Nuclear Power Station about questions relating to the basic concept, the proposed upgrading measures, and conclusions to be drawn from the GRS report.

In addition, three statements by the Kurchatov Institute were delivered together with the comments. They are contained in Annex A.4 to this report.

The Soviet side did not express any comments about the main part of the safety assessment (Sections 1 to 9). Some individual opinions, suggestions, and supplements to Annex A.3 are contained in this comment. There is agreement between the two parties about all items not addressed in the Soviet comment.

10.2 Comment on Annex A.3 by the Main Designer and the Scientific Leader

10.2.1 Comment Regarding Upgrading Measures

Ad A.3.1.1 (Materials)

Ad 1. The material of the reactor pressure vessel ensures a service life in accordance with the design, provided the water in the accumulators and storage tanks of the emergency cooling system is heated to a temperature of 55 °C. The neutron flux acting on the walls of the reactor pressure vessel can be reduced in the following ways:

- Use of shielding assemblies.

- Choice of a nuclear fuel core loading with a low neutron leakage (this would require additional neutron physics calculations to be performed).

Ad 2. The ALÜS leakage detection system made by Siemens must be installed, or corresponding system to be developed by the Soviet Union by 1993 must be used.

Ad 5. During inspection, the bottom weld is not accessible. The upper weld can be inspected if the protective core cover is removed and subsequently fitted back in position.
We believe that such inspections are not necessary because stress calculations and positive operating experience have demonstrated the life of the collectors, including the welds. During manufacturing the collector welds were tested and proved to be 100% undistructed. In addition, the upper weld is tested indirectly during operation.

Ad 6. The excess weld metal will be removed during assembling and commissioning according to a technology developed by the component manufacturer and main designer.

Ad 7. It is possible to exchange the feedwater supply lines inside the installed system generators for lines made of austentic material at accessible locations of potentially endangered areas.

Ad 9. An early intervention of the plant personnel is necessary not only in cases of leaks of steam generator tubes but also in all cases of leaks inside the steam generator between the primary and the secondary system as well as in some other accidents.

Ad A.3.1.2 (Process Engineering)

Ad 27. At present, the main coolant pumps are replaced in USSR nuclear power plants with reactors of the type W-213. The cooling and lubrication of the radial-axial pump bearing is performed by water.

Ad 3.1.4 (Instrumentation and Control)

Ad 1. Agreement is expressed with the introduction of these additional reactor scram criteria. However, attention is drawn to these items:

- The instrumentation and control equipment for introducing the DNB shutdown criterion is made available by the German side.

- The "pressurizer filling level high" shutdown criterion must be connected with the "pressure in the primary system low" criterion by a logical "AND."

Ad 2. The "pressurizer filling level high in connection with primary systems pressure low" reactor scram signal meets the requirement (according to the Soviet comment on Item 1) of diverse reactor scram initiation for the "pressurizer safety valves stuck open" accident.

Ad 5. For reactors of the type W-213, the AKNP-2 neutron flux measuring system is to be exchanged for the AKNP-7 system. The requirement to automate neutron flux measurement will then be met (except for automatic adjustment of the neutron flux >110% of the permissible reactor power setting).

Ad 6. Introducing the "drop of filling level below the design value by L = 400 mm" reactor scram criterion in only one steam generator is not meaningful. The existence of a sufficient water reserve in the steam generator has been demonstrated in calculations for design basis accidents, provided the 2-out-of-6 logic has been realized. If this filling level criterion (L = 400 mm) is realized in the of 1-out-of-6 steam generators logic, there could be erroneous activation as a result of controller failures and measurement failures in the electronic system.

Ad 8.1 The introduction of an "activity increase in the main steam system" reactor scram criterion is suggested which, among other things, would also be effective in case of a heater tube break (Item 1 from A.3.1.4.).

Ad 9. Manual measures are to be taken for accident management in the first thirty minutes after the initial event in all cases of leakages from the primary system into the secondary system.

Ad 23. At present, such a filling level probe for backfitting the type W-213 reactor is under development.

10.2.2 Comment on Analyses and Verifications

Ad A.3.2.1 (Materials)

Ad 2. The Soviet side is prepared to support the German side in procuring additional documents from the manufacturer. For this purpose, letters will be written by the Soviet side to the plant manufacturer.

Ad 4. There is an analysis of the load acting on the nozzle connections as a result of a slant of the reactor pressure vessel of approx. 4mm.

Ad 13. Studies of cold strands for reactors of the type W-213 exist not only in the Loviisa Nuclear Power Station but also in the Kola Nuclear Power Station. With respect to thermal shock loads acting on the reactor pressure vessel (cold water strands), the calculations of the resistance to brittle fracture were made completely in accordance with Soviet standards.

Ad A.3.2.2 (Process Engineering)

Ad 3. The exchange cross bores in the fuel assemblies are located below and above the fuel rod sections. They serve to decrease the pressure differentials in loss-of-coolant accidents and, hence, to relieve the fuel assemblies. They have no influence on cross mixing among the fuel rods. Consequently, the same subchannel factors, K_q and $K_{\Delta H}$, are assumed as for fuel assemblies without exchange cross bores.

Ad 4. An explanation is included in Annex A.4.

Ad 6. The sufficiency of the design of the emergency cooling systems has been verified in calculations for the design basis accidents.

Ad 7. The activity limits of the water in the primary system are contained in the catalog for fuel assembly equipment of the WWER-440 in accordance with Soviet standards.

Ad 10. The "break of a tube in the steam generator" accident was calculated without taking into account the loss of off-site power case. Personnel will have to intervene in order to manage the accident (localization and isolation of the leakage).

Ad 11. At the time of project development, the "collector break on the primary side" accident was not considered to be a design basis accident. In accordance with the current rules, this accident has to be analyzed like other accidents exceeding the design basis. This may require specific technical and administrative measures of risk minimization to be derived and defined.

Ad 12. and 13.

The Soviet side is convinced of the usefulness of such accident analyses, but points out that no non-steady state three-dimensional computer codes are at present available for design basis calculations. The existing codes have not been verified sufficiently well.

Ad 14. As far as variations in location and size of leaks in a main steam line are concerned, the following calculations underly the project:

- Leakage in the main steam line within the confinement

- Leakage in the main steam line outside the confinement

- Leakage in the main steam collector

- Opening and subsequent failure to close of the steam generator safety valves or the atmospheric main steam dump station (BRU-A)

In the calculations referred to above, the reactor inlet temperature is determined and the effectiveness of existing interlocks and automatic systems is demonstrated by evidence.

Ad 6., 9., 14., 18.

> In 1990 all accident analyses for the type WWER-440/W-213 have been updated on the basis of the requirements of the Soviet accident guidelines. A joint seminar is recommenced to discuss these accident analyses.

Ad 28. The verification of a reliable functioning of the reactor scram system is available up to RPV slant of 3 mm.

Ad A.3.2.4 (Instrumentation and Control)

Ad 1. A setting value between 20 and 100 kPa/s has been established in the project for the "main steam collector break" accident.

> More precise calculations indicate the usefulness of a setting value in the range between 40 and 50 kPa/s. No additional studies are necessary.

10.2.3 Comment on the Documentation

Ad A.3.3.1 (Materials)

Ad 1., 3., 4.

> SU-A ad 1., 3., and 4.
> It is appropriate to convene a trilateral seminar (Federal Republic of Germany, France, USSR) on the service life of components in nuclear power plants of the type WWER-440/W-213 in order to discuss the following questions:
>
> - Establishing characteristic data of loads for strength calculations, standards on which to base strength, and considering plant operation data.
>
> - Materials fatigue must be taken into account and analyzed, including the methods of calculating stress conditions.
>
> - Methods of component surveillance in operation, including monitoring end of service life.

Ad 2. The Soviet side can elaborate such a status report.

Ad 5. and 6.

> The desired information can be given by the manufacturer with the participation of the main designer.

Ad 9. There is a possibility to inspect all steam generator tubes from the inside, i.e. from the side of the primary system, by using the "System Interkontrol" eddy current inspection unit. In the Soviet Union, this materials testing unit is currently being tried on the steam generators of the WWER-1000 and can also be used on the WWER-440/W-213 steam generators.

10.3 Comment on Annex A.3 by the General Project Engineer

10.3.1 Comment on Upgrading Measures

Ad A.3.1.1 (Materials)

Ad 8. The project documentation about the installation of the resin catchers was delivered to the purchaser of the nuclear power plant for implementation.

Ad Item A.3.1.2 (Process Engineering)

Ad 10. Administrative measures are outlined in the operating manual to cope with accidents with leakages of one heater tube. They take into account that the primary isolating valves will not close completely. For this reason and also with a view to normal operation, additional automatic measures are not considered to be useful. In case of several heater tube leakages or collector leakage administrative-technical measures must be provided in order to prevent inadmissable discharges from the affected steam generator.

Ad 14. The safety valves are designed to protect the heat exchangers from overpressure. At present, technical measures are being developed in the USSR to protect heat exchangers from overpressure and to prevent activity releases in the case of pipe breaks, e.g. by installing rupture disks to protect the heat exchanger of the component cooling system for the main coolant pumps. Analogous measures are designated for the component cooling system SUS.

Ad 20. Emergency power can be supplied to the startup and shutdown pumps if the results of the load accommodation of the Diesel generators (emergency power balance) permit (see conclusions on Items 9. and 10. in Section 1.3.1.3).

Ad 22. In the opinion of the Soviet side, work must be speeded up to implement the "leak-before-break" principle and introduce modern diagnosing systems for monitoring the materials of pipes and equipment.

Ad 29. In accordance with the design, the confinement isolation valves for the normal operating systems are supposed to open after the blockage preventing these valves from opening has been canceled automatically when the pressure in the steam generator box has dropped to atmospheric pressure.

Ad A.3.1.3 (Electrical Engineering)

Ad 9. and 10.

In accidents, the safety of operation of the unit is ensured by the three-channel safety system which is supplied with emergency power. To increase the reliability of the drives of normal operating systems important to the functioning of the main equipment and to the availability of the nuclear power plant, it is advisable to provide for an additional electric power supply system.

Ad A.3.1.4 (Instrumentation and Control)

Ad 9. To prevent operators' errors in an accident sequence, active components of the safety systems are blocked against being shut off. These blocks can be canceled only by measurement signals indicating the safe status of the reactor plant. In major accidents with leakages from the primary system into the secondary system, no administrative and technical measures are provided to introduce temporary blockages of operator actions to control the safety systems (see conclusion on Item 12 in Section A.3.1.2).

Ad 17. This requirement is realized in USSR nuclear power plants.

Ad 31. and 32.

> The problems referred to were discussed in the commissioning phase of Unit 5. Proposals were elaborated to eliminate these deficiencies, and measures for realization were defined (installation of filling level sensors with signalling to the control room of the unit).

Notes on A.3.1.3 (Electrical Engineering) and A.3.1.4 (Instrumentation and Control)

> Both sides state that the replacement of the instrumentation and control systems, the upgrading of the station service power supply system, and the implementation of process engineering measures requires further cooperation. This cooperation is useful, especially in linking instrumentation and control systems to process engineering systems, in the installation of the equipment so that its influence on building structures is taken into account, and in harmonizing the algorithms for the modes of operation of the automatic systems (control circuits, open control loops).

Ad A.3.1.5 (Civil Engineering Aspects)

Ad 3. According to information provided by the purchaser of the plant, the seismicity on site is 4 points on the MSK-64 scale, which does not require any additional measures to be taken according to Soviet rules.

Ad 4. and 8.

> Analyses conducted by the Atomenergoprojekt Institute show that it is not possible to insure separation of the pipes and equipment on the 14.7 m platform so as to prevent mutual interference, unless additional measures are taken to accommodate accident loads in the building structure. Proposals are also contained in the conclusion on Item 22 in Section A.3.1.2.

10.3.2 Comment on Analyses and Verifications

Ad A.3.2.3 (Electrical Engineering)

Ad 4. and 5.

On the basis of calculations performed, measures have been defined jointly with the architect engineer of Unit 5:

- Changes in the setting marks

- Installation of additional automatic systems

- Rerouting of cables.

10.4 Conclusions

The position adopted by the experts from the Federal Republic of Germany in choosing the most important upgrading measures for Unit 5 of the Greifswald Nuclear Power Station agrees basically with the approaches and opinions expressed by the Soviet experts.

Not all questions addressed in the safety assessment by GRS have been taken up in the present Soviet comments. Further meetings have been scheduled at which these questions will be settled and the necessary agreement about technical points should be achieved.

The Soviet experts indicated that the following upgrading measures are being envisaged for type W-213 nuclear power plants in the Soviet Union:

- Installation of a passive system to prevent explosible concentrations of hydrogen in the pressurized compartments

- Replacement of the detachable part of the main coolant pumps by a part with water cooling and water lubrication of the radial axial pump bearing, and with an encapsulated oil lubrication system for the bearings of the electric drive

- Passive system for residual heat removal in case of complete failure of the electric power supply including uninterrupted station service power supply

- Passive system for pressure reduction and filtering accident-induced discharges from the pressurized compartments
- Changing the oil supply to the turbo generators to non-burnable lubrication oil (OMTI).

Annex A.1

Nuclear Power Plants of the type WWER-440/W-213

Location	Country	Unit	Commissioning date	1)
Kola	USSR	3	03/81	
		4	10/84	
Rovensk (Rovno)	USSR	1	12/81	
		2	12/81	
Greifswald	Germany	5	03/89	2)
		6	under construction since '80	3)
		7	under construction since '81	3)
		8	under construction since '81	3)
Zarnowiec	Poland	1	under construction since '82	4)
		2	under construction since '82	4)
		3	under construction since '88	4)
		4	under construction since '88	4)
Bohunice	CSFR	3	08/84	
		4	08/85	
Dukovany	CSFR	1	02/85	
		2	01/86	
		3	11/86	
		4	06/87	
Mochovce	CSFR	1	under construction since '83	
		2	under construction since '83	
		3	under construction since '84	
		4	under construction since '84	
Paks	Hungary	1	12/82	
		2	09/84	
		3	09/86	
		4	08/87	
Juragua	Cuba	1	under construction, commissioning planned 1993	5)
	W-318	2	under construction, commissioning planned 1996	5)
Loviisa	Finland	1	02/77	5)
		2	11/80	5)

Notes:

1) First synchronization of grid

2) The commissioning phase was interrupted in November 1989

3) Construction work discontinued for the time being

4) Construction work stopped

5) Modified plants, e.g. with a containment

Annex A.2

Participating Firms and Institutions

The following firms and institutions were commissioned by GRS to work on subproblems within this study:

- Hosser, Haß + Partner,
  Ingenieurgesellschaft für Bauwesen und Brandschutz mbH,
  Braunschweig

- Prof. Dr.-Ing. Josef Eibl,
  Beratender Ingenieur für Bauwesen,
  Karlsruhe

- Rheinisch-Westfälischer Technischer Überwachungs-Verein e.V.,
  Essen

- Staatliche Materialprüfungsanstalt (MPA)
  Universität Stuttgart,
  Stuttgart

- Technischer Überwachungs-Verein Bayern e.V.,
  München

- Technischer Überwachungs-Verein
  Norddeutschland e.V.,
  Hamburg

Annex A.3

Summary of the Upgrading Measures Derived from the Studies, and of the Analyses and Documents Needed for further Investigations

This Annex A.3 contains a complete list of the upgrading measures derived from the studies, of the analyses and verifications needed for further investigations, and of the documents and data still required. The list, in addition, features the comments, suggestions and additions made in the Soviet statement (Section 10) on specific points, some of which have been incorporated into the recommendations. Excerpts from the comments by the main designer and scientific leader (Section 10.2) are marked SU-A, those by the general project engineer (Section 10.3) are marked SU-B. Mutual agreement exists on all points about which no statements were made by the Soviet side.

A.3.1 Upgrading Measures

A.3.1.1 Materials

(1) The EOL fluence at the wall of the reactor pressure vessel must be limited (from Sec. 4.2.2).

SU-A

The material of the reactor pressure vessel ensures a service life in accordance with the design, provided the water in the accumulators and storage tanks of the emergency cooling system is heated to a temperature of 55 °C. The neutron flux acting on the reactor pressure vessel wall can be reduced in the following ways:

- *Use of shielding assemblies.*

- *Choice of a nuclear fuel core loading with a low neutron leakage (this would require additional neutron physics calculations to be performed).*

(2) Special leak monitoring systems must be provided for leak detection in the RPV top nozzles (from Sec. 4.2.2).

SU-A

The ALÜS leakage detection system made by Siemens must be installed or a corresponding system to be developed in the Soviet Union by 1993 must be used.

(3) Accessibility for in-service inspection of the main coolant pipe and the connecting pipes of the pressurizer must be improved (from Sec. 4.2.2).

(4) Roots not penetration-welded, e.g. at spray nozzles and heater elements of the pressurizer and at NB 500 thermal buffers at the main gate valve, must be eliminated (from Sec.4.2.2).

(5) At the steam generator collector, inspection possibilities for the weld surfaces on the secondary side must be provided for the bottom junction welds (from Sec. 4.2.2).

SU-A

During inspection, the bottom weld is not accessible. The upper weld can be inspected if the protective core cover is removed and subsequently fitted back in position.
We believe that such inspections are not necessary because stress calculations and positive operating experience have demonstraded the life of the collectors, including the welds. (from Sec. 4.2.2)

(6) Existing restrictions of non-destructive tests resulting from the test geometry and from excess weld metal respectively must be eliminated (from Sec. 4.2.2).

SU-A

The excess weld metal will be removed during assembling and commissioning according to a technology developed by the component manufacturer and main designer.

(7) The pressurized area of the feedwater system must be protected against erosion-corrosion by suitable substitutions of materials or by claddings (from Sec. 4.2.2).

SU-A

It is possible to exchange the feedwater supply lines inside the installed steam generators for lines made of austentic material at accessible locations of potentially endangered areas. (from Sec. 4.2.2)

(8) Resin catchers must be installed downstream of the special water treatment system (SWA 1 and SWA 1a) (from Sec. 4.2.2).

SU-B

The project documentation about the installation of the resin catchers was delivered to the purchaser of the nuclear power plant for implementation.

(9) An operational system must be installed for automatic monitoring of the water chemistry parameters in the primary and secondary systems (from Sec. 4.2.2).

SU-A

An early intervention of the plant personnel is necessary not only in cases of leaks of steam generator tubes but also in all cases of leaks inside the steam generator between the primary and secondary system as well as in some other accidents (from Sec. 4.2.2).

(10) The compressed-air tanks and pipes of the Diesel startup air system must be made of corrosion-resisting steel (from Sec. 8.2.1).

(11) Development and application of suitable materials testing techniques for quality assurance of the passive mechanical components installed (from Sec. 8.2.1).

A.3.1.2 Process Engineering

(1) A redundant fast boron injection system supplied with emergency power must be established for reactor scramming with an injection pressure above the operating pressure (from Sec. 4.1.1, Sec. 6.1.3.10).

SU-B

(2) The LP emergency coolers are cooled directly with sea water (absence of an activity barrier; contamination hazard of the emergency coolers for long-term heat removal). In addition, the bearings of the three high-pressure emergency cooling pumps are cooled in the single-leg component cooling system (NKW-B). Manual switching to the component cooling system of the main coolant pumps is possible in principle, but hardly feasible in an accident. Backfitting a three-leg component cooling system is necessary (from Sec. 6.1.2.1, 6.1.2.2).

(3) Operating the residual heat removal system requires two series-connected isolating valves to be opened. To ensure reliable opening, another valve group should be installed parallel to the existing isolating valves. The group of valves must be monitored for internal leakage (from Sec. 6.1.2.1).

(4) The reliability of feeding from the accumulators into the reactor must be improved. The existing switching system must be modified. The possibility must be analyzed of the accumulator shutoff ball failing, or one of the two isolating valves wrongly closing (from Sec. 6.1.2.1).

(5) To prevent overpressurization of the accumulators, the tightness of the two check valves in the accumulator connecting pipes must be monitored (from Sec.6.1.2.1).

(6) The reliability of the position indication of the shutoff balls in the accumulators must be improved. Otherwise there could be danger of not recognizing an inadvertant closure of the accumulators (from Sec. 6.1.2.1).

(7) In each pump discharge line of the emergency cooling systems a motor-driven valve must open when actuated. These motor-driven valves must be replaced by check valves with monitoring capability of the pipe section between the valves. Additional isolating valves must be provided for repair cases (from Sec. 6.1.2.1).

(8) In case of a sump return flow plugging up, the flow of water to the other two sumps must be ensured (e.g., interconnection of the three building sumps) (from Sec. 6.1.2.2)

(9) Should it become necessary to heat the boric acid solution in the storage tank in order to ensure resistance to brittle fracture, it has to be studied whether recooling of the minimum water discharge of the high-pressure emergency cooling pumps must be achieved through the component cooling system to be newly installed (from Sec. 6.1.2.2).

(10) Accidents with leaks from the primary system into the secondary system (e.g. leak in the steam generator collector) must be demonstrated to remain within the accident planning levels. Possibilities should be examined, and, if necessary, ensured to improve tight closing of the main coolant lines by the main gate valves, without subsequent manual retightening, as well as reliable closing of the main gate valves under the full differential pressure (from Sec. 6.1.2.6).

SU-B

Administrative measures are outlined in the operating manual to cope with accidents with leakages of one heater tube. They take into account that the primary isolating valves will not close completely. For this reason, and also with a view to normal operation, additional automatic measures are not considered to be useful. In case of several heater tube leakages or collector leakage administrative-technical measures must be provided in order to prevent inadmissable discharge from the affected steam generator.

(11) In case of leakage in a steam generator heater tube or in a collector and failure of a main gate valve, overfeeding of the defective steam generator by the high-pressure emergency cooling pumps must be prevented by automatic measures (from Sec. 6.1.2.6).

(12) A pressurizer relief valve with preclosing capability must be backfitted. The actuation pressure must be set below the response pressure of the fluid-actuated steam control valves of the safety valves (from Sec. 6.1.2.4).

(13) The rupture disk in the pressurizer relief tank must be constructed and installed so that it will respond reliably only after the actuation pressure has been reached (from Sec. 8.2.1).

(14) The safety valves on the component cooling water side of the heat exchangers must be designed at least to the break of one heat exchanger tube. In case of an activity increase on the compartment cooling water side the confinement isolation valves and the isolation valves on the primary side of the heat exchangers must be closed automatically (from Sec. 6.1.2.7).

SU-B

The safety valves are designed to protect the heat exchangers from overpressure. At present, technical measures are being developed in the USSR to protect heat exchangers from overpressure and to prevent activity releases in the case of pipe breaks, e.g. by installing rupture disks to protect the heat exchanger of the component cooling system for the main coolant pumps. Analogous measures are designated for the component cooling system SUS.

(15) The activation of the steam generator safety valves must be made redundant, and the safe opening and subsequent, reliable closing functions must be verified (including a reliability analysis) (from Sec. 6.1).

(16) A controllable steam generator safety valve with a lower actuating pressure and equipped with an isolating valve, and an isolating valve upstream of the atmospheric main steam dump (BRU-A), should be installed; 100% steam dumping capacity must be ensured through safety valves which cannot be isolated (from Sec. 6.1.3.4).

(17) Enhancing the reliability and, if necessary, exchanging of all components of the atmospheric main steam dump station (BRU-A) and the bypass stations, the valves important for pressure protection of the steam generators, and the valves in the feedwater and condensate systems (from Sec. 8.2.1).

(18) An independent emergency feedwater system must be installed. This system must be protected against spreading in-plant flooding (from Sec. 6.1.3.6).

(19) The present emergency feedwater system must be connected to the feedwater tanks (from Sec. 6.1.3.6).

(20) There is no automatic start of the startup and shutdown pump when the filling level in the steam generators drops; the pump is not connected to the emergency power system. This deficiency must be eliminated in connection with a new feedwater concept (from Sec. 6.1.3.2).

SU-B
Emergency power can be supplied to the startup and shutdown pumps if the results of the load accommodation for the Diesel generators permit this (emergency power balance) (see conclusions to Items 9 and 10 in Sec. A.3.1.3).

(21) Additional possibilities of feeding emergency feedwater must be created (e.g. connecting nozzle for accident management measures) (from Sec.6.1.3.6).

(22) Technical solutions must be elaborated to prevent consequential damage arising from a leak in the region of the 14.7 m platform to other pipes or pieces of equipment in adjacent areas of the compartments (from Sec. 6.1.3.7, 8.2.4).

SU-B
In the opinion of the Soviet side, work must be speeded up to implement the "leak-before-break" principle and modern diagnosing systems introduced for monitoring the materials of pipes and equipment.

(23) Additional analyses are required of the concept of feedwater control from the LP preheaters up to the steam generators so as to optimize control of the feedwater system (from Sec. 8.2.5).

(24) Motor-driven isolating valves with position indicators in the main control room must be installed in the feedwater intake pipes instead of the planned isolation by blanks (from Sec. 6.1.3.7).

(25) The safety valves of the cooldown system must be designed for water blowdown (from Sec. 6.1.3.1).

(26) Oil leaks occur in the area of the main coolant pumps. The halon extinguishers used in this area are not able to protect the compartments. Effective extinguishing systems must be installed (from Sec. 7.1.2.1).

(27) Fundamental revision of the main coolant pump and its oil circuit so as to eliminate oil leakages (from Sec. 8.2.1 and 6.1.4).

SU-A

At present the main coolant pumps are replaced in USSR nuclear power plants with reactors of the type W-213. The cooling and lubrication of the radial-axial pump bearing is performed by water.

(28) The <NB 80 pipes in the confinement must be run as in the project design (from Sec. 6.1.2.3).

(29) In case of an accident, it should be possible to reopen several confinement isolation valves after the confinement has been isolated and a leakage detected. This concerns e.g. valves in the feed pipe of the volume control system (feed system) which would open up an additional feeding possibility in the emergency cooling case (from Sec. 6.1.2.3).

SU-B

In accordance with the design, the confinement isolation valves for the normal plant operating systems are supposed to open after the blockage preventing these valves from opening has been canceled automatically when the pressure in the steam generator box has dropped to atmospheric pressure.

(30) Doubling the discharge area through the check valves into the air traps of the pool-type pressure suppression system (from Sec. 5.2.7).

(31) Preventing consequential damage caused by jet forces, reaction forces, missiles, thermal loads, and moisture in the pool-type pressure suppression system (from Sec. 5.2.7).

(32) Installing a leak extraction system at all penetrations and all leaks detected (from Sec. 5.2.7).

(33) Improved leakage seals at the air locks (from Sec. 5.2.7).

(34) Modifications in building structures, such as moving the staircase at the lock platform (Room G202A) into an area with a lower local dose rate, and modifications of the doors as well as changes to the lock systems, of the transfer lock and the emergency locks respectively. In this connection, it should be examined whether the requirement to improve the radiation protection of the staff and modernize the lock can be met by a new building (from Sec. 7.3.2).

(35) The nitrogen system must be checked for the possible impact of the medium in the primary systems and, if necessary, must be upgraded (from Sec. 8.2.1).

(36) Fire protection dampers exist in the ventilation systems at only a few points. Backfitting is required in areas with safety-related separations and in staircases (from Sec. 7.1.2.1).

A.3.1.3 Electrical Engineering

(1) Improving the grid side of the station service power supply (from Sec. 6.2).

(2) Due to deficiencies in the automatic transfer system for the standby grid feeding capability, a revision must be made in which the existing voltage and current conditions are taken into account (from Sec. 6.2).

(3) Changing the power supply of the emergency cooling chain in such a way that, upon actuation, it will be switched to the emergency power Diesel units only when the station service power supply system fails (from Sec. 6.2 and 6.1.2.1).

(4) Establishing an uninterruptible possibility to switch from the emergency power system back to the station service system when the grid voltage returns (from Sec. 6.2).

(5) Conceptual changes in the uninterruptible power supply system, e.g.:

- Separate rectifiers and inverters (from Sec. 6.2, 8.2.3).

- Double feeding of the DC-loads or DC-distribution systems (from Sec. 8.2.3).

- Using two batteries per system channel (from Sec. 6.2).

(6) Improving the reliability of the Diesel generators, including the startup air supply, the Diesel fuel injection system, and a revision of the Diesel control system, to prevent damage due to overloading from arising under any operating condition, even in case of maloperation (from Sec. 8.2.1).

(7) Increasing battery capacity. According to an RSK requirement, the discharge time of 30 minutes must be increased to 2-3 hours (from Sec. 6.2).

(8) The physical separation of emergency power generation and distribution systems must be observed on secondary routes and on the cable gallery underneath the main control room (from Sec. 6.2).

(9) Emergency power supply must be provided to the feed pumps of the volume control system (from Sec. 6.1.2.3).

(10) The cooldown system must be supplied with emergency power (from Sec. 6.1.3.1).

SU-B, ad 9. and 10.
In accidents, the safety of operation of the unit is ensured by the three-channel safety system which is supplied with emergency power. To increase the reliability of the drives of normal plant operating systems important to the functioning of the main equipment and to the availability of the nuclear power plant, it is advisable to provide for an additional electric power supply system.

(11) The connection of the main coolant pumps (MCP) to the 6kV standby lines must be designed so that there can be no simultaneous failure of more than 3 MCP as a result of a short circuit in the bus (from Sec. 6.2).

(12) Checking the design of the station service system and the electrical protection systems in such a way that defects in the areas of the transformers, switching

systems and distribution networks, and loads can be detected and shut off reliably (from Sec. 8.2.3).

A.3.1.4 Instrumentation and Control

(1)   Introduction of the following reactor scram initiation criteria:

- Increase in main steam activity

- DNB ratio low

- Pressure in the primary system high

- Pressurizer filling level high

(from Sec. 6.1.2.6, 6.1.3.2, 6.3).

SU-A
Agreement is expressed with the introduction of these additional reactor scram criteria. However, attention is drawn to these items:

- *The instrumentation and control equipment for introducing the DNB shutdown criterion is made available by the German side.*

- *The "pressurizer filling level high" shutdown criterion must be connected with the "pressure in the primary system low" criterion by a logic "AND."*

(2)   For the "pressurizer safety valves stuck open" accident, the diverse actuation for the "opening of a pressurizer safety valve" reactor scram must be backfitted (from Sec. 6.1.2.4, 6.3).

SU-A
The "pressurizer filling level high in connection with primary systems pressure low" reactor scram signal meets the requirement (according to the Soviet comment on Item 1) of diverse reactor scram initiation for the "pressurizer safety valves stuck open" accident.

(3) Reliable activation of reactor scram prior to the actuation of the rupture membrane between the steam generator box and the pool-type pressure suppression system shaft (from Sec. 5.2.7).

(4) The "failure of the last operating turboset" criterion for actuating reactor scram can be rendered inoperative by an easily accessible switch. This interlock must be fully automated (from Sec. 6.1.3.3 and 8.2.2).

(5) In the neutron flux measurement system, adjustments of the measuring chambers, changes in measuring range, and power matching of the "neutron flux >110% of the permissible reactor power" reactor scram criterion must be automated (from Sec. 6.1.3.9 and 8.2.2).

SU-A

For reactors of the type W-213, the AKNP-2 neutron flux measuring system is to be exchanged against the AKNP-7 system. The requirement to automate neutron flux measurement will then be met (except for automatic adjustment of the neutron flux >110% of the permissible reactor power setting).

(6) In order to protect the secondary-side water supplies, it is recommended to initiate reactor scram already at a level dropping to LDE < -400 mm in only one steam generator (e.g. in the case of a break of a feedwater pipe) (from Sec. 5.1.2).

SU-A

Introducing a "drop of filling level below the design value around L = 400 mm" reactor scram criterion in only one steam generator is not meaningful. The existance of sufficient water reserve in the steam generator has been demonstrated in calculations for design basis accidents provided the 2-out-of-6 logic has been realized. If this filling level criterion (L = 400 mm) were realized in the 1-out-of-6 steam generators logic, there could be erroneous activation as a result of controller failures and measurement failures in the electronic system.

(7) Introduction of a turbine power limitation or reactor scram as a function of the number of failed main feedwater pumps (from Sec. 6.1.3.2, 5.1.2).

(8) Eliminating the deficiencies in automatic limits and protective actions where these features are missing, for instance:

- Rapid shutdown of the unit through the secondary system (e.g. in case of a leak in a heater tube)

- Ensuring sufficient shutdown reactivity under all operating conditions (from Sec. 5.1.1 and 6.1.3.9, 6.1.2.6, 6.3).

SU-A, ad 8.1
The introduction of an "activity increase in the main steam system" reactor scram criterion is suggested, which, among other things, would be effective also in case of a heater tube break (Item 1 from A.3.1.4).

(9) Measures must be introduced which allow the 30-minute rule to be observed (after the onset of an accident, no manual measures are required within 30 minutes) also in the "steam generator heater tube leak" accident, among others (from Sec. 5.1 and 6.3).

SU-A
Manual measures of accident management are required within the first 30 minutes after the initial event in all cases of leakages from the primary system into the secondary system.

SU-B
To prevent operators' errors in an accident sequence, active components of the safety systems are blocked against being shut off. These blocks can be canceled only by measurement signals indicating the safe state of the reactor plant. In accidents with major leakages from the primary system into the secondary system, no administrative and technical measures are provided to introduce temporary blockages of operator actions to control the safety systems (see conclusion on Item 12 in Section A.3.1.2).

(10) Creating a diverse activation to recognize the loss of off-site power case (underfrequency), and resetting the activation level initiating the undervoltage case (from Sec. 6.2 and 6.3).

(11) Introduction of diverse signal processing in the measurement and control trains up to the actuating relay. This does not apply to the pulse tubes from the reactor measurement nozzle (from Sec. 6.3).

(12) Self-monitoring, testability, and fault-tolerant design in all areas (from Sec. 6.3, 8.2.2).

(13) Use of automatic control for systems to avoid incorrect operation (e.g. feedwater supply, condensate pumping, main coolant pump) (from Sec. 8.2.2).

(14) Automatic comparison of the readings from multichannel measurements and automatic signaling in case of inadmissible deviations; avoidance of different signal statuses in the control room and in the interlocking logic (from Sec. 8.2.2).

(15) The switching status and the setpoint vs. actual value comparison of the interlocking positions must be monitored automatically (from Sec. 6.1.3.9).

(16) Improving signalization in the control room for accidents and wrong setting of valves in safety systems (monitoring normal operational position) (from Sec. 8.2.2).

(17) Improving torque activation and deactivation of selected valve drives (from Sec. 8.2.2).

SU-B
This requirement is realized in USSR nuclear power plants.

(18) A position indicator for the valve in the high-pressure preheater bypass line must be installed in the main control room so that there is a possibility to check the feedwater supply in case of failure of the high-pressure pre-heater column (from Sec. 6.1.3.6 and 6.2.3.2).

(19) Improving the quality of the measurements of boric acid concentration (from Sec. 8.2.2).

(20) Improving computer reliability for logging the main parameters in the course of disturbances; if necessary, replacement of existing computers (from Sec. 8.2.2).

(21) Installation of instruments resistant to accidents (from Sec. 6.3.).

(22) Improving accident monitoring instrumentation (from Sec. 6.3).

(23) Installation of a filling level probe in the reactor pressure vessel (from Sec. 6.3.).

SU-A
At present, such a filling level probe for backfitting the type W-213 reactor is under development (from Sec. 6.3)

(24) Eliminating a number of points where redundant installations of the instrumentation and control system are located in the same fire area (from Sec. 6.3).

(25) The functions of the main control room and the standby control room are not isolated completely. Such isolation is necessary for reasons of fire protection (from Sec. 7.1.2.1).

(26) The control rooms must be reconstructed fundamentally with regard to ergonomic aspects (from Sec. 6.4 and 6.3).

(27) Reliable monitoring of the operating condition of the main coolant pumps by suitable criteria, such as rotational speed (from Sec. 8.2.2).

(28) Reliable monitoring of safety-related filling levels, such as pressurizer, steam generator, accumulator (from Sec. 8.2.2).

(29) Improving leak detection (from Sec. 8.2.2).

(30) Automatic monitoring of the sumps in the reactor building with signaling to the control room (hazards reporting system) (from Sec. 8.2.2).

(31)   The redundancy and reliability of the water level sensors in the three pump compartments of the emergency core cooling system and residual heat removal systems must be improved (from Sec. 6.1).

SU-B, ad 31. and 32.
The problems referred to were discussed in the commissioning phase of Unit 5. Proposals were elaborated to eliminate these deficiencies, and measures for realization were defined (installation of filling level sensors signaling to the control room of the unit).

(32)   The monitoring capability for the isolation of the building drainage system connecting the pump compartments, which does not exist at present, must be backfitted (from Sec. 6.1.2.1, 8.2.4).

(33)   The entire instrumentation and control system must be exchanged (from Sec. 6.3, 8.2).

(34)   As the fire alarm system is made up of various sub-systems of which not all are properly harmonized, the basic concept and the sensors need to be checked (from Sec. 7.1.2.1).

SU-B
Both sides state that the replacement of the instrumentation and control systems, the upgrading of the station service power supply system, and the implementation of process engineering measures require further cooperation. This cooperation is useful, especially in linking instrumentation and control systems to process engineering systems, in the installation of equipment so that its influence on building structures is taken into account, and in harmonizing the algorithms for the modes of operation of the automatic systems (control circuits, open control loops).

A.3.1.5 Civil Engineering Aspects

(1) To avoid sequences of events spreading beyond redundancies as a result of major leakages of the main cooling water system, of the service water system and of emergency cooling water, structural evidence or measures are required to eliminate the deficiencies resulting from the lack or insufficient verification of physical separation. Cross connections spreading beyond redundancies through drainage systems or pipes of the filling system must be secured in the closed position during normal operation. This applies to the following building areas with safety-related systems:

- Intake structure for the service water system A:
  service water pumps

- Turbine hall:
  emergency feed pumps and feedwater pumps

- Reactor building:
  HP and LP emergency cooling systems and sprinkler system
  Pit cooling system

(from Sec. 7.1.2.2 and 6.1.3.8).

(2) Ensuring and demonstrating sufficient separation of the Units 5 and 6 in such a way that inadmissible mutual interference (e.g. through sumps of the common special building) is avoided (from Sec. 8.2.4).

(3) Some building structures need to be upgraded in some parts for the "earthquake" load case. In the German codes and standards (KTA 2201.1), safety-related plants must be so designed that their safety-related function is preserved in the design basis earthquake. These plants must be designed to a horizontal acceleration on the free building surface of at least 0.8 m/s² (from Sec. 7.2).

SU-B
According to information provided by the purchaser of the plant, the seismicity on site is 4 points on the MSK-64 scale, which does not require any additional measures to be taken according to Soviet rules.

(4) Arranging the turbines in the turbine hall parallel to the reactor building implies the risk of missiles being launched against the reactor building in case of a turbine explosion; however, this danger can be reduced by structural protective measures (from Sec. 7.2).

SU-B, ad 4. and 8.
Analyses conducted by the Atomenergoprojekt Institute show that it is not possible to ensure separation of the pipes and equipment on the 14.7 m platform so as to prevent mutual interference, unless additional measures are taken to accommodate accident loads in the building structure. Proposals are also contained in the conclusion on Item 22 in Section A.3.1.2.

(5) For reasons of conventional building codes, measures must be taken in the turbine hall to prevent the generation and spreading of large area fires (e.g. encapsulation of the main fire loads), so that the necessary special permits may be granted with respect to excessive sizes of fire compartment areas and lengths of escape routes. In addition, systematic checks, especially of the escape routes, must also be carried out in other areas of the plant (from Sec. 7.1.2.1).

(6) The deficiencies in fire protection, as far as the reactor building is concerned, result mainly from the co-location of cables of various redundancies of safety-related systems. The three redundancies must be separated for purposes of fire protection (from Sec. 7.1.2.1).

(7) Some of the cable compartments, cable coatings with insulating substances, and fire protection doors are of insufficient quality; backfitting measures are necessary (from Sec. 7.1.2.1).

(8) Damage to instrumentation and control installations located in compartments below feedwater and main steam lines (14.7 m platform) must be prevented (from Sec. 6.3).

A.3.1.6 Administration and Operations Management

(1) Automatic fuses must not be short-circuited for test purposes (from Sec. 8.2.2).

(2) Improving the job order and isolation concept for better coordination of activities of shift personnel, radiation protection, fire protection, maintenance, and technical departments (from Sec. 8.2.5).

(3) Clear definition of competences between the architect engineer and the power plant personnel (from Sec. 8.2.5).

(4) Establishing a revision concept to ensure nuclear safety (from Sec. 8.2.5).

(5) Definition of the nominal set points of special valves and switches in the operating manual for various operating states; if necessary, technical designs must be adapted (from Sec. 8.2.5).

(6) Improving quality assurance, from component manufacturing up to installation (from Sec. 8.2.5).

(7) Comprehensive tests of the interaction of systems in the non-nuclear commissioning phase; reliable removal of temporary commissioning setups (from Sec. 8.2.5).

(8) Regular checking of measuring stations at sufficiently short intervals, and regular checking of drainage sumps by visual inspection (from Sec. 8.2.5).

(9) Cranes must not pass over important safety-related systems while the plant is in the power operation, if no technical measures of protection are possible (from Sec. 8.2.5).

(10) Revising the mode of operation of the laboratory (rules for performing analyses, organization) (from Sec. 8.2.5).

(11) Revision of present operational radiation protection regulations for adaptation to the criteria in the Radiation Protection Ordinance and in applicable codes and regulations; in addition, revising

- the organization,

- competences, and

- duties of radiation protection

in the light of an appropriate quality assurance system so as to eliminate all present administrative deficits (from Sec. 7.3.2).

(12) Reduction of radiation exposure by increased mechanization and the use of remote handling techniques, especially for work on the entire primary system, including the reactor pressure vessel, pressurizer and steam generator, and, if necessary, for testing and surveillance work in the steam generator box. In this respect, the additional detailed measures already identified by the operator in the Units 1-4 to reduce radiation exposure must also be implemented (from Sec. 7.3.1).

(13) Use of a high-quality respiration protection gear (from Sec. 7.3.1).

(14) Use of a suitable directly readable personnel dose monitoring system for occupationally exposed personnel. The system must not only provide dose monitoring capability but must also have capabilities of dose warning, monitoring of the access control area, and electronic data evaluation and data processing. Moreover, radiation protection surveillance of accessible rooms and areas must be improved (from Sec. 7.3.2).

A.3.2 Analyses and Verifications

A.3.2.1 Materials

(1) The verifications of the stability of primary systems components and their supports must be repeated with present methods of calculation and, if

necessary, supplemented by FEM calculations for specific load cases (from Sec. 4.2.2).

(2) To evaluate the quality assurance concept employed by component manufacturers, the information available must be supplemented by inspections of the documentation of checks made during construction which are kept by the manufacturers. Non-destructive tests of the base materials must be verified and, if necessary, repeated (from Sec. 4.2.2).

SU-A
The Soviet side is prepared to support the German side in procuring additional documents from the manufacturers. For this purpose, letters will be written by the Soviet side to the plant manufacturers.

(3) The results of the basic ultrasonic inspection of the base metal areas (rings and bottom) of the reactor pressure vessel must be presented, including the nozzles of NB 250 (from Sec. 4.2.2).

(4) The consequences arising from the measured slant of the reactor pressure vessel with respect to loads acting on the nozzle connections must be evaluated (from Sec. 4.2.2).

SU-A
There is an analysis of the load acting on the nozzle connections as a result of a slant of the reactor pressure vessel of approx. 4 mm.

(5) Ultrasonic inspection of the steam generators must be repeated completely in accordance with the criteria set forth in the KTA Codes and Standards (from Sec. 4.2.2).

(6) The defect in Unit 2 which developed in 1982 from a blind hole in the steam generator collector flange must be examined for any conclusions resulting with respect to the operation of Unit 5 (from Sec. 4.2.2).

(7) The validity of the radiographic inspections of the welds of the main coolant pumps and main gate valves must be determined. If necessary,

supplementary inspections with optimized inspection techniques must be carried out (test specimens) (from Sec. 4.2.2).

(8) In the pressurizer, the welds of the vessel must again be inspected ultrasonically for longitudinal and transverse flaws (from Sec. 4.2.2.).

(9) Surface crack tests must be made up for the welds of the vessels and pipes of the secondary system (from Sec. 4.2.2).

(10) It must be investigated to what extent leak cross sections larger than 80 cm² can be excluded in the steam generator collector and whether the leak-before-break criterion applies (from Sec. 6.1.2.6).

(11) Checking the stud bolts of the main coolant pump when replacing a gasket (materials inspection), and the bearing connections and throttles; if necessary, replacement by a new design (from Sec. 8.2.1).

(12) The use of sufficiently leaktight condensers with chrome nickel steel or titanium pipes must be checked as a precondition for changing to the high-AVT mode of operation in the secondary system (from Sec. 4.2.2).

(13) Analyses of the formation of cold water strands should be conducted similar to the PTS Study drafted in Finland for the Loviisa Nuclear Power Station (from Sec. 5.1.3).

SU-A
Studies of cold strands for reactors of the type W-213 exist not only in the Loviisa Nuclear Power Station, but also in the Kola Nuclear Power Station. With respect to thermal shock loads acting on the reactor pressure vessel (cold water strands), the calculations of resistance to brittle fracture were made completely in compliance with Soviet standards

(14) Verifying the integrity of the plastic deflection hoods in the pool-type pressure suppression system under accident conditions, with aging taken into account (from Sec. 5.2.7).

(15) Verification of the accommodation capability of dynamic loads in dampers, pools and building structures during condensation process in the pressure suppression pools (from Sec. 5.2.7).

A.3.2.2 Process Engineering

(1) The accident analyses confirm that the rate at which control assemblies are inserted is enough when the first activation criterion is effective. It remains to be examined whether the reactor is shut down reliably in all design basis accidents, also if only the second activation criterion is effective, i.e. if the first activation is assumed to fail (from Sec. 4.1.1).

(2) Sticking of the most effective control rod is taken into account in the design of the reactor scram system. The Accident Guidelines require verification of the sufficient improbability of operating transients in an assumed failure or partial failure of the reactor scram system. It needs to be verified whether the reactor scram system is sufficiently reliable (from Sec. 4.1.1).

(3) For thermohydraulic design, the subchannel factors for the increase in enthalpy, $K_{\Delta H}$, and the heat flux density, K_q, must be verified for the current assembly design with exchange cross bores (from Sec. 4.1.2).

SU-A
The exchange cross bores in the fuel assemblies are located below and above the fuel rod sections. They serve to decrease the pressure differentials during loss-of-coolant accidents and, hence, to relieve the fuel assemblies. They have no influence on cross mixing among the fuel rods. Consequently, the same subchannel factors, K_q and $K_{\Delta H}$, are assumed as for fuel assemblies without exchange cross bores.

(4) The observance of the minimum permissible DNB values must be verified for the design basis transients with regard to the effective power limitation (HS-4 or HS-3. This also requires detailed information about the accuracy of the DNB correlations used (from Sec. 4.1.2).

SU-A

An explanation is included in Annex A.4.

(5) Power density data must be included in the automatic power limitation and the reactor scram capabilities respectively. The algorithms used for power density monitoring must be verified (from Sec. 4.1.2).

(6) To demonstrate the sufficiency of the design of the emergency cooling systems, an analysis of the entire "double-ended break of the main coolant pipe" accident sequence, including an analysis of the extent of damage, is necessary (from Sec. 5.1.1).

SU-A

The sufficiency of the design of the emergency cooling systems has been demonstrated in calculations for the design basis accidents.

(7) The activity level of the coolant in the primary system must be determined from the Soviet limits indicated for fuel rod failures (from Sec. 4.1.3).

SU-A

The activity limits of the water in the primary system are contained in the catalog for the fuel assembly equipment of the WWER-440 in accordance with the Soviet standard.

(8) An analysis must be carried out of the break of an accumulator feed pipe leading into the annulus of the pressure vessel (from Sec. 5.1.1).

(9) Accident analyses must be conducted of the small leak with a leak size equivalent to NB 113, and of the break of the pipe connecting the pressurizer and the safety valves (from Sec. 5.1.1).

(10) Detailed analyses must be carried out of the "double-ended break of a steam generator heater tube" accident. In particular, those analyses must be conducted from which automatic measures are derived to prevent inadmissible activity releases to the outside. These analyses must include the assumption, among others, that the main isolating valves on the primary side

do not close completely. In addition, variants must be studied with and without the occurrence of the emergency power case (from Sec. 5.1.1).

SU-A
The "break of a tube in the steam generator" accident was calculated without taking into account the loss of off-site power case. To manage the accident, activities by the personnel are required (localization and isolation of the leakage).

(11) Detailed analyses must be carried out of the break of the steam generator collector to demonstrate the effectiveness of suitable upgrading measures (from Sec. 5.1.1).

SU-A
At the time of project development, the "collector break on the primary side" accident was not considered to be a design basis accident. In accordance with current rules, this accident has to be analyzed like other accidents exceeding the design basis. This may require specific technical and administrative measures of risk minimization to be derived and defined.

(12) Supplementary analyses of the ejection of control elements must be carried out with a three-dimensional (3D) reactor dynamics code. This applies in particular to analyses of the ejection of decentralized control elements (from Sec. 5.1.2).

(13) Supplementary analyses with 3D core models must be conducted of the reactivity feedback associated with a leak in the main steam system (from Sec. 5.1.2).

SU-A, ad 12. and 13.
The Soviet side is convinced of the usefulness of such accident analyses but points out that no non-steady state three-dimensional computer codes are at present available for design basis calculations. The existing codes have not been verified sufficiently well.

(14) Supplementary analyses are required of the break and leakage respectively of a main steam pipe, in which the water entrainment phenomena on the secondary side are modeled as realistically as possible. The locations and sizes of the leaks must be varied systematically so that the most adverse impacts on the core inlet temperatures and the effectiveness of different reactor protection criteria (releasing interlocks HS 4 to HS 1) can be determined. In case the basic safety of the pipes on the 14.7 m platform cannot be confirmed, analyses must be performed of the breaks of several main steam pipes (from Sec. 5.1.2).

SU-A

As far as variations in location and size of leaks in a main steam line are concerned, the following calculations underly the project:

- *Leakage in the main steam line within the confinement*

- *Leakage in the main steam line outside the confinement*

- *Leakage in the main steam collector*

- *Opening and subsequent failure to close of the steam generator safety valves or the atmospheric main steam dump station (BRU-A)*

In the calculations referred to above, the reactor inlet temperature is determined and the effectiveness of existing interlocks and automatic systems is demonstrated.

(15) No analyses are available of operating transients associated with failures of the reactor scram system (ATWS). ATWS analyses are demanded in the RSK Guidelines for some selected operational transients (from Sec. 5.1.2).

(16) Detailed investigations must be carried out of the pressure buildup and pressure differentials in the confinement (from Sec. 5.2.7).

(17) The local arrangement of pressure sensors must be checked for reliable pressure measurement in the confinement (from Sec. 6.1.2.4).

(18)   Detailled investigations must be carried out into the performance of the sprinkler system considering failure criteria (from Sec. 5.2.7).

SU-A, ad 6., 9., 14., and 18.
In 1990 all accident analyses for the type WWER-440/W-213 have been updated on the basis of the requirements of the Soviet accident guidelines. A joint seminar is recommended to discuss these accident analyses.

(19)   In an emergency, using the sprinkler pumps to replace failed low-pressure emergency cooling pumps for residual heat removal is considered to be useful. The reliability of possible technical solutions must be investigated (from Sec. 6.1.2.1).

(20)   If the connecting pipe between the hot and the cold legs of the primary recirculation pipe is necessary to prevent water plugs (which still remains to be investigated), the valves in the connecting pipe should be permanently open. No valves are foreseen for the connecting pipes in Units 7 and 8 (from Sec. 6.1.2.1).

(21)   The functioning capability of the pressurizer safety valves must be demonstrated for the passage of steam-water mixtures and water (from Sec. 6.1.2.4).

(22)   The pressure buildup in the pool-type pressure suppression system and in the confinement must be determined for the "pressurizer safety valves stuck open" accident. If necessary, pressure sensors must be backfitted in the shaft of the pool-type pressure suppression system (from Sec. 6.1.2.4).

(23)   It must be investigated whether the pipes, building isolation valves, and the pipes between the confinement and the building isolation valves have been designed to the primary system pressure (from Sec. 6.1.2.7).

(24)   The high-pressure feed pipes of the post-accident cooling system and the feed pipes of the volume control system have no whipping restraints. The possibility of consequential failures arising from pipe leaks must be examined (from Sec. 6.1.2.2).

(25) The ferritic purge lines and emergency feedwater lines have no pipe whip restraints even within the confinement. It must be examined whether pipe whip restraints are required (from Sec. 6.1.3.6).

(26) The possibility of a leak in a connecting pipe of the primary system outside the confinement causing consequential damage to confinement isolation valves and pipes must be investigated (from Sec. 6.1.2.7).

(27) Verification of the secondary system's sufficient protection against entry of sea water (especially in the area of the shutdown condensers) (from Sec. 8.2.1).

(28) Possible influences of the control rod insertion time resulting from the slant of the RPV has to be checked again (the slant is about 1.5 mm)

SU-A
The verification of a reliable function of the reactor scram system is available up to RPV slant of 3 mm.

A.3.2.3 Electrical Engineering

(1) Checking the emergency power balance and, if necessary, increasing the Diesel generator capacity (from Sec. 6.2).

(2) The electric power supply to the main isolating valves (HAS) comes from the emergency power Diesel system in cases of emergency cooling and loss of off-site power. In establishing the emergency power balance, the power consumption of the motors driving the main gate valves must be taken into account (from Sec. 6.1.2.6).

SU-B
See Item 10 in Section A.3.1.2.

(3) Verification of the suitability of the cables run into the confinement (from Sec. 8.2.2).

(4) Demonstration of the suitability of all switching systems of safety-related loads at all voltage levels; where necessary, exchanging switching systems for those whose suitability has been verified (from Sec. 8.2.3).

(5) Verifiable calculations of the maximum and minimum short circuit currents, and determination of the minimum voltage limits for all loads (from Sec. 8.2.3).

SU-B, ad 4. and 5.
On the basis of calculations performed, measures have been defined jointly with the purchaser of Unit 5:

- *Changing the setting marks*

- *Installation of additional automatic systems*

- *Rerouting of cables.*

(6) Demonstration of sufficient reliability of the installations for uninterrupted electric power supply; separation of the functions for battery charging and supplying the safe main distribution board (from Sec. 8.2.3).

A.3.2.4 Instrumentation and Control

(1) It is not apparent from the analyses of the break of the main steam collector whether, and how reliably, the "exceeding the pressure drop rate of 80 kPa for at least 5 s" reactor scram criterion is met. Additional studies must be conducted before a suitable criterion can be defined (from Sec. 5.1.2 and 6.1.3.5).

SU-A
A setting value between 20 and 100 kPa/s is foreseen in the project for the main steam collector break accident. More precise calculations indicate the usefulness of a setting value in the range between 40 and 50 kPa/s. No further investigations are necessary.

(2) Verifying the reliability of electrical contacts, e.g. for the control and protection system drives (from Sec. 8.2.2).

(3) Checking the concept of power supply for safety-related measurements and interlocks, including the electrical protection systems, on the basis of a failure effect analysis in order to prevent failures of protective actions or actuations of protective actions whose impact on safety is not evident, and also to avoid other erroneous actuations, e.g. demeshing of the safety systems (from Sec. 8.2.2).

A.3.2.5 Civil Engineering Aspects

(1) The accommodation of loads arising from leakage accidents in the turbine hall must be verified in a supplementary procedure for the penetrations of the main steam and feedwater pipes in wall C (from Sec. 4.2.2).

(2) Whether there is any possible danger of the intake structure and the pump building being flooded can only be determined after the probability of occurrence of floods have become known. If necessary, special protective measures must be taken (from Sec. 7.1.3).

(3) In the light of site-specific conditions and probabilistic considerations it needs to be investigated whether or to what extent events are important which arise from an airplane crash, external blast waves from chemical reactions, and external impacts of hazardous substances (from Sec. 7.1.3 and 7.2).

A.3.2.6 Administration

(1) Demonstration of sufficient qualification of the personnel (from Sec. 8.2.5).

A.3.3 Documentation and Data

A.3.3.1 Materials

(1) The calculations performed by the plant manufacturer of loads arising in the RPV internals during normal operation and loss-of-coolant accidents must be supplied (from Sec. 4.1.3).

SU-A ad 1., 3., and 4.
It is appropriate to convene a trilateral seminar (Federal Republic of Germany,

France, USSR) on the service life of components in nuclear power plants of the type WWER-440/W-213 in order to discuss the folloing questions:

- *Establishing characteristic data of loads for strength calculations, standards on which to base strength, and considering plant operations data*

- *Materials fatigue must be taken into account and analyzed, including the methods of calculating stress conditions*

- *Methods of component surveillance in operation, including monitoring end of service life.*

(2) A status report must be elaborated which represents the current state of knowledge about the reliability of processing and the behavior under neutron exposure and corrosion characteristics of the 15Ch2MFA reactor pressure vessel steel (from Sec. 4.2.2).

SU-A
The Soviet side can elaborate such a status report.

(3) The characteristic mechanical and engineering data contained in the certificates cannot be allocated to any specimen location. In particular, the shapes of specimens used for the toughness tests have not been indicated. Supplementary information is required in this area (from Sec. 4.2.2).

(4) Some numerical information about the mechanical and engineering data and chemical analyses respectively which differ greatly from the specifications must be checked. In addition, differences in the toughness data must be cleared up which may have arisen from conversions of dimensions (from Sec. 4.2.2).

(5) More detailed information is required about the test procedure of austenitic-ferritic welds, especially in the reactor pressure vessel forged-on nozzle necks (from Sec. 4.2.2).

(6) A inspection program of the nozzles and the area of holes in the reactor pressure vessel top must be presented (ultrasonic inspection from the inside,

inspection by means of a TV camera from the inside and the outside)(from Sec. 4.2.2).

SU-A, ad 5. and 6.
The desired information can be given by the manufacturer with the participation of the main designer.

(7) A mechanized internal inspection (ultrasound, visual inspection) must be provided for in-service checks of the main coolant line and the connecting pipes of the pressurizer. The method of testing the longitudinal welds of bends must be upgraded (from Sec. 4.2.2).

(8) A method of inspecting austenitic-ferritic welds must be upgraded (from Sec. 4.2.2).

(9) A concept of inspection of the steam generator tubes must be presented which also includes the areas of the bends (from Sec. 4.2.2).

SU-A
There is a possibility to inspect all steam generator tubes from the inside, i.e. from the side of the primary system, by using the "System Interkontrol" eddy current inspection unit. In the Soviet Union, this materials testing unit is currently being tried out on the steam generators of the WWER-1000 and can also be used on the WWER-440/W-213 steam generators.

A.3.3.2 Civil Engineering Aspects

(1) Complete and testable verifications of stress analyses must be presented (from Sec. 7.2).

(2) A "general approval by the building supervision authorities" or an "individual approval" is required for the composite steel cell type of construction (from Sec. 7.2).

(3) The following are the documents to be examined:

- Design and anchorage of the removable parts

- Supports and anchorage points of the components

- Jet forces in case of an accident

- Quality assurance.

A.3.3.3 Administration

(1) Compiling an operating manual in line with the requirements contained in the operating manuals of nuclear power stations in the Federal Republic of Germany (from Sec. 8.2.5).

(2) In the revision of the operating manuals, the procedures for startup and shutdown must be described more precisely (from Sec. 6.1.3.9).

(3) Establishing of a reliable central modification service for the plant documentation (from Sec. 8.2.5).

(4) Matching the plant documentation to the actual plant structure (from Sec. 8.2.5).

Annex A.4

Analyses of Fuel Rod Behavior by the Kurchatov Institute

A.4.1   Calculating of the Critical Heat Flux of Fuel Rods in type WWER Reactors

The critical heat flux is calculated by the formula,

$$q_{crit} = 0.795 \, (1-x) \, \exp(0.105 \, p - 0.5) \cdot (\rho w) \, \exp(0.184 - 0.331 \, x) \cdot (1 - 0.0185 \, p) \quad (1)$$

with

q_{crit} - critical heat flux in MW/m²

x - relative enthalpy at the location of boiling

ρw - mass velocity in kg/m²s

p - pressure in MPa.

The formula was set up on the basis of experimental data derived from tests with seven rod bundles and uniform heat flux over the length of the fuel rod. A lattice with a pitch of 12.2 to 12.75 mm was used for spacing the fuel rods. The axial distance between lattices is 255 mm. The heated rod length was 1.75 - 3.5 m. In addition to the experimental data from OKB Gidropress, data from the Kurchatov Institute for 7-, 19-, and 37-rod bundles were also used in statistical error evaluation. A total of 776 experimental points were employed. The formula describes all 776 points with a root mean square error of $\sigma = 13.1\%$ and a deviation from the mean of $\mu = 1.01$. The range of applicability of the formula is as follows:

Pressure	7.45-16.7 MPa
Mass velocity	700-3800 kg/m²s
Relative enthalpy at the location of boiling	-0.07 to +0.4

The nonuniformity of the heat flux over the length of the fuel rod is described by a correction factor in the formula:

$$q_{crit}^{nonuniform} = q_{crit}^{uniform} \cdot F \qquad (2)$$

with $\quad F = \{[q(z) \cdot l]^{-1} \cdot \int_{z-1}^{z} q(z)dz\}^n$

and $\quad n = 3.79 - 19{,}61 [p/p_{crit}] + 17{,}88 [p/p_{crit}]^2$

and

$\quad\quad p_{crit} = 22.13$ MPa

$\quad\quad l = 55\, d_t$ relaxation length, m

$\quad\quad d_t =$ thermal diameter, m

The correction factor was determined on the basis of 438 experimental points which were determined in rod bundles with nonuniform heat flux over the length. The most important information derived from these studies is contained in two papers:

1. Experimental Investigations and Statistical Data Analysis of Boiling in Rod Bundles for WWER Reactors.
   Authors:　　J.A. Besrukov, W.I. Astachov et al.,
   　　　　　　Teplonergetika No. 2, 1976

2. Studies of the Influence of the Axial Heat Flux Profile of the Fuel Rod upon Boiling in a Rod Bundle.
   Authors:　　W.I. Astachov, J.A Besrukov et al.,
   　　　　　　Publication by the CMEA Country Seminar on Thermal Physics TF-78, Budapest, 1987, pp 589-600

A.4.2 Studies of Fuel Rod Behavior in Accidents

A.4.2.1 Software Covering Fuel Rod Behavior in Accidents

The FRASM-PC code is used for calculations of the thermodynamic state of fuel rods in accidents. This completed and accelerated code has been developed on the basis of the Czech FRAS code /1, 2/. The code is a modular computation program. In line with licensing requirements, the code contains a number of blocks taking into account the multifaceted physical processes going on in fuel rods during accidents. The FRASM-PC code has been planned for analytical calculations of the thermomechanical state of the fuel rods of WWER-type power reactors during accidents. The processes going on are calculated individually in the modules of the code. The logical structure of the code allows the individual computations to be combined to a uniform complex. The modular design of the code also allows calculations to be performed permanently and quickly both with the entire code and with individual modules. In addition, this code structure permits computations to be simplified, or made more complicated, by adding other modules, in line with the problem at hand. The subroutine library describing the properties of the N-1 alloy /3/ can be connected to the FRASM-PC code. These subroutines are necessary to determine the changing physico-mechanical properties of the fuel cladding during an accident. This code module is based on a large number of experimental data obtained under laboratory conditions. The subroutine library uses one out of ten modules containing various experimental data about the high-temperature properties of fuel assembly claddings, including those made of a zirconium alloy. This allows comparative calculations to be run with different experimental data, but also with foreign codes, and permits the code to be verified on the basis of different experiments. The structural terms in the code module contain data about corrosion and the criteria of cladding destruction. The reason is the importance of the processes described by these modules. The first module calculates the corrosion of the fuel rod cladding at elevated temperatures on the inner and outer surfaces. The third module determines the failure criteria of the cladding in accidents. With these models, the power of the exothermal oxidation reaction and the oxidation geometry of the fuel rod cladding is determined; both factors are taken into account in the other modules. Oxidation of the inner surface is calculated after the cladding has begun to leak. All modules describing the properties of the cladding are based on a series of experimental data determined on specimens of an N-1 alloy.

A.4.2.2 Code Verification by Experimental Data

The code was verified by published data derived from experiments performed in special test rigs with nuclear heating. For comparison with data calculated by means of this code experimental data were used as published in the paper /4/. They were employed to verify the code for accident calculations with ramp-type reactivity rises. The results obtained with the code were in satisfactory agreement with the experimental findings. The second experiment used to verify the FRASM-PC code represents the simulation of a LOCA. The MT-1 experiment was run in the NRU reactor to verify the FRAP-T6 code /5/. In this case, the thermomechanical behavior of the fuel rod materials was studied in the presence of pressure drops and flow reductions and in the rewetting phase. For code verification, the measured fuel rod cladding temperature was used as a basis, and a comparison was made of the calculated and the experimental data about deformation and the time of failure. The calculated results correlate sufficiently well with the experimental findings. The results are indicated in paper /6/.

A.4.2.3 Calculated Analysis of Fuel Rod Behavior in WWER during Accidents

The code was elaborated on the basis of experiments performed to determine the influence of the initial thermomechanical state of WWER-440 fuel rods upon fuel rod behavior in a maximum design basis accident - break of the NB 500 main coolant pipe at the inlet of the reactor - for various phases of operation (irradiation histories of the fuel rods), taking into account the variance in the characteristic data of the fuel rods with respect to the onset of irradiation. The data on which the calculation was based were test results of neutron physics and thermohydraulic calculations of the break of a main coolant pipe at the inlet of the WWER-440 reactor. The axial temperature distribution over the fuel rod cladding was assumed conservatively for the most highly loaded fuel rod, with rewetting taken into account, in order to analyze in detail the thermomechanical condition of the fuel rod. Calculations by means of the PINO4-M code /7/ were used to obtain the initial data for the thermomechanical characteristic curves of the fuel rods. That program was employed to perform calculations of fuel rods for quasi-steady state reactor operation. The results of the calculations show the internal gas pressure to dominate the behavior of the fuel rods. The internal gas pressure is determined not only by the initial condition of the fuel rods, but also by stress induced deformation and by thermal conditions during the accident. The variances in the initial plant data and in burnup do not greatly detract from the safety of WWER-440 reactors, but add to the uncertainty of fuel rod behavior in loss-of-coolant accidents. The results are indicated in paper /8/.

A.4.2.4 Experimental Data on the Physico-Mechanical Properties of Fuel Rod Materials under Normal and Accident Conditions

All physico-mechanical properties of the fuel rod materials were determined under special laboratory conditions in the temperature range between 20 and 1200 °C for the fuel assembly cladding and, up to operating temperature, for the fuel. The properties of the cladding under normal conditions were examined in specimens irradiated with a neutron fluence of 10^{17} neutrons/cm² up to 10^{21} neutrons/cm² with irradiation temperature steps up to 350 °C. All the major characteristic curves of the physico-mechanical properties of fuel rod claddings are listed in paper /3/ and include the following characteristics:

- Thermal physics constants (temperatures of phase transitions, specific heat of melting, melting temperature);
- Specific enthalpy;
- Coefficient of thermal conductivity;
- Diametral and radial thermal expansion;
- Axial thermal expansion;
- Toughness modulus according to Junk;
- Poisson coefficient;
- Strength coefficients;
- Calculations of realistic deformation diagrams on the basis of given deformations or stresses in the plastic range;
- Limits to mechanical properties;
- Anisotropy factors of elasticity;
- Radiation-induced swelling in various directions, with the texture coefficients taken into account.

The stress deformation state of fuel rod claddings in accidents is modeled on the basis of data describing the mechanical properties obtained when studying high-temperature creeping of unirradiated specimens under laboratory conditions or dummy fuel rods in special rigs under oxidizing and inert atmospheres. These data are used to obtain the functional interdependencies of creep rates, creep and deformation rates in the high-temperature regime for various methods:

- A function in the form of Norton's law;

- Theory of tough plastic deformation based on Levi's and v. Mises' theory of creeping;

- Special equation describing the mechanical state (the FRASM-PC code is used):

$$\varepsilon = A(T) \sinh [B(T) \cdot (\sigma - \sigma_0)]$$

with

ε - rate of plastic deformation,

σ - stress,

σ_0 – resilience,

T - temperature,

A,B - coefficients.

All these models can be used to calculate the deformations of cladding tubes during accidents. Analogous data were obtained about the functions of cladding oxidation in the high-temperature range and on the destruction criterion. The codes make use of all properties modeling the thermomechanical condition and fuel rod behavior under various operating conditions and during accidents of WWER reactors.

References, Annex A.4.2

/1/   F. Pazdera:

FRAS Program of Thermomechanical Calculations of Fuel Rod Behavior in WWER Plants under Accident Conditions and for Verification Calculations. Problems of Atomic Science and Technology, "Materials Behavior in Nuclear Technology" Series, Edition No. 2 (27), 1988

/2/   F. Pazdera, M. Valach, V. Vrtilkova:

Studies of the Behavior of Light Water Reactor Fuel Rods under Accident Conditions

IAEA-TC-579/24, Vienna, November 10-13, 1986

/3/   B.J. Wolkov, V.F. Viktorov, P.A. Platanov, A.W. Rjasanzeva:

Subroutine Library on Physicomechanical Fuel Rod Cladding Properties Made of the N-1 Alloy

Report No. 4941 by the Kurchatov Institute, 1989

/4/   N. Onishi, K. Ishijima, S. Tanzawa:

Report about Supercooled Film Boiling Heat Transfer under Reactivity Accident Conditions in Light Water Reactors

N. S. E. 1984, vol. 88, pp. 331-41

/5/   T. Vanderkaa:

FRAP-T6, an Independent Code for MT-1 LOCA Simulation Tests in the NRU Reactor

/6/   B.J. Wolkov, F. Pazdera, M. Valach, N.B. Sokolov, J. Linek:

Comparison of the FRAS, RAPTA, SSYST-3 Computer Codes with Results of the MT-1 Reactor Experiment Recalculation of the Mechanical Fuel Rod Condition. CSFR-USSR

Bilateral Seminar, December 1990

/7/   P.N. Strijov, V.V. Yakovlev, F. Pazdera:

Improved Version of the PIN Code and its Verification

Preston, England, September 19-22, 1988, IAEA-TC-657/3.4

/8/ B.J. Wolkov, W.W. Yakovlev:
Calculations of the Influence of Indeterminate Initial Values upon the Thermomechanical Fuel Rod Behavior in WWER Reactors in Loss-of-Coolant Accidents.
British-Soviet Seminar, April-May, 1990

A.4.3. Modeling Fuel Rod Behavior in WWER Reactors under Normal Operating Conditions

A.4.3.1 Description of the PIN-Micro Code

The PIN-Micro code, which is installed on an IBM PC-AT, is used to model fuel rod behavior in WWER reactors under normal operating conditions (quasi-steady state operation) /1-3/. Figure 1 shows the interconnection, as used in the PIN-Micro code, of processes going on in fuel rods of the type WWER. Specifically for this code, a subroutine package was elaborated which describes the characteristics of the N-1 alloy. That alloy is the cladding tube material in fuel rods of the WWER-440. The program package contains simple functional dependencies for creep behavior (radiation-induced and thermally induced) and for radiation-induced cladding growth. These dependencies were obtained from analyses of the data derived from in-pile tests and post-irradiation examinations in the MR reactor and the hot cells of the Kurchatov Institute respectively. The creep behavior of the cladding was studied under steady-state and non-steady state conditions. Descriptions of some models can be found in paper [2]. The realistic irradiation history of the fuel rods is modeled under the assumptions of constant power, coolant temperature, and fast neutron flux for one time step for calculations run by means of the PIN-Micro code. The following data are determined for each time step:

- Radial temperature distribution in the fuel and in the cladding;

- Width of the gap or contact pressure between the fuel and the cladding;

- Changes in fuel dimensions as a result of thermal expansion, cracking, leaking, and swelling;

- Changes in the dimensions of the cladding tube as a result of thermal elasticity, creeping, and radiation induced growth;

- Generation and release of gaseous fission products into the free volume of the fuel rod;

- Modification of the fuel structure as a result of the formation of equi-axial oblong grain and the formation and/or enlargement of the central bore.

Moreover, the axial extension of the fuel column and the fuel pellets as well as the inner gas pressure is calculated for all fuel rods. To calculate the fuel column data mentioned above, the fuel column is subdivided into axial segments of the same length (a maximum of 20), with each segment in turn being subdivided into concentric radial rings (a maximum of 50). In each score (control volume), all characteristic data of the fuel rod are assumed to be constant. The fuel rod cladding is regarded as only one score in the radial direction, i.e. the thin-cladding hypothesis is used. For a discrete computation scheme for the fields of temperature, deformation of the fuel pellets and other characteristic data, the code contains some nested cycles and iteration loops. In the innermost iteration loop, the conductivity of the gap, the temperature fields in the fuel, the radial displacement of the fuel and of the cladding are calculated. This is followed by a calculation, in segments, in the axial direction. This cycle includes iteration loops ensuring convergence with respect to the gas composition and pressure within the fuel rod cladding. The outermost cycle is a time cycle modeling the power history in steps. The PIN-Micro code consequently is an integral, full-scale, axisymmetrical, thermomechanical code. With respect to its structure, the PIN-Micro code is a quasi-two-dimensional or, more precisely, a 1.5-dimensional code. This means that the fields of temperature, deformation, gas releases, gap conductivity etc. are calculated independently for each axial segment by a one-dimensional solution approach. The different segments are connected in the calculation of the extension of the fuel rod column and the pellets, the compensation volume, the gas composition, and the pressure under the fuel rod cladding. The composition of the gaseous fission products released from the axial segments and the internal gas pressure are assumed to be equal for all axial segments of the entire fuel rod within one time step.

A.4.3.2 Verification of the PIN-Micro Code

The results obtained with the PIN-Micro computer code were verified by comparison with a few integral fuel rod data obtained in the appropriate in-pile and post-irradiation examinations carried out by the Kurchatov Institute. For verification, data were used which originated in thoroughly studied fuel rods equipped with in-core sensors. The main characteristic data of the fuel rods for which the comparison was made between calculated and experimental data are the temperature in the fuel center, the internal gas pressure, the length extension of the fuel and of the fuel pellets. These data were measured during irradiation. In the verification of the PIN-Micro code, especially data from the Soviet-Finnish SOFIT experimental program /1/ were used. To verify the code at high burnups, post-irradiation examination data were used, among them studies of full-scale fuel rods which had been in Unit 4 of the Novovoronesh Nuclear Power Station over three fueling periods. For these data, agreement with the PIN-Micro code was found to be good. This demonstrates its applicability to modeling the fuel rod behavior in WWER reactors under normal operating conditions. More information about the verification results can be found in papers /2,3/. As an example describing the use of the PIN-Micro code, the next section will contain the results of calculations performed to optimize the internal gas pressure in fuel rods for WWER-440 reactors.

A.4.3.3 Optimization of the Initial Gas Pressure in Fuel Rods for WWER-440 Reactors of Increased Power

It is well known that a certain amount of overpressure improves the thermophysical characteristics of fuel rods. A number of variants were calculated at various pressures to optimize the initial gas pressure for fuel rods of WWER-440 reactors of increased power. The optimization was performed in the light of the magnitude of the final internal pressure, which turned out to be an important parameter in the analysis of accidents. Diagram 2 shows the results of this optimization. It can be seen that the best value of the initial gas pressure for fuel rods of this type is 5 - 7 bar. Increasing the pressure still further will not improve the thermophysical characteristics of the fuel rods.

References, Annex A.4.3

/1/ Strijov, P., et al.:
Research of WWER-440-type Fuel Rods in MR Reactor, IAEA International Symposium on Improvements in Water Reactor Fuel Technology and Utilization, Stockholm, Sweden, September 15-18, 1986

/2/ Strijov, P., et al.:
An Improved Version of the PIN Code and its Verification, IAEA Technical Committee Meeting on Water Reactor Fuel Element Computer Modeling in Steady State, Transient and Accident Conditions.
Preston, England, September 19-22, 1988

/3/ Strijov ,P., et al.:
Computer and Experimental WWER Fuel Rod Modeling for Extended Burnup, IAEA Technical Committee Meeting on Fuel Performance at High Burnup for Water Reactors, Studsvik, Sweden, June 5-8, 1990

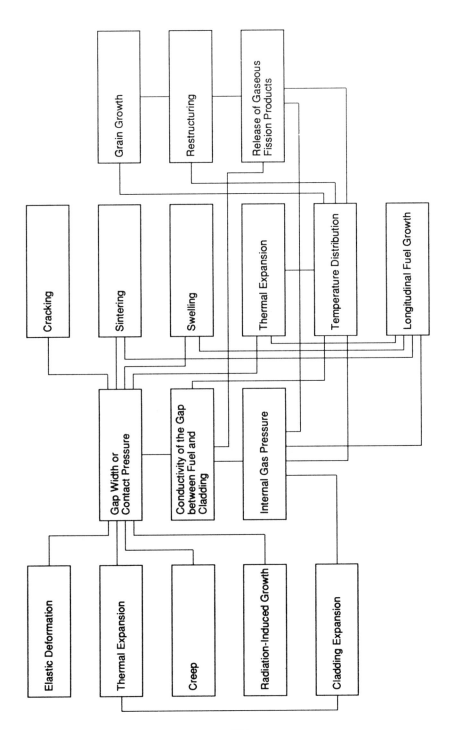

Diagram 1: Schematic diagram of processes in fuel rods of WWER reactors

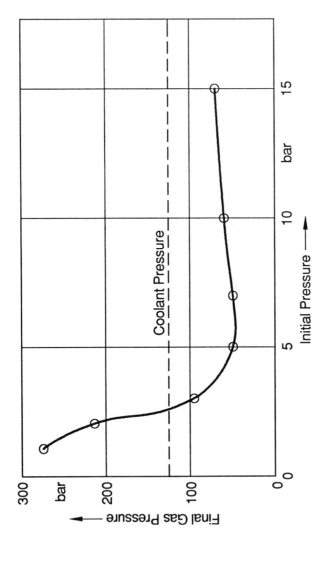

Diagram 2: Optimizing the initial pressure in fuel rods of WWER-440